THE POLITICS OF OFFSHORE OIL

Edited by
Joan Goldstein

Foreword by Senator Bill Bradley

PRAEGER SPECIAL STUDIES • PRAEGER SCIENTIFIC

Library of Congress Cataloging in Publication Data
Main entry under title:

The Politics of offshore oil.

Includes bibliographical references.

Contents: Introduction / by Joan Goldstein—California,
threatening the Golden Shore / Elizabeth R. Kaplan—The
search for an ocean management policy, the George   Bank
case / Charles S. Colgan—[etc.].
   1. Offshore oil industry—Government policy—United
States—Addresses, essays, lectures.   2. Offshore oil
industry—Environmental aspects—United States—Addresses,
essays, lectures.   3. Coastal zone management—United
States—Addresses, essays, lectures.
I. Goldstein, Joan.

HD9566.P64 1982   333.8′232′0973   82-7697
ISBN 0-03-059813-3   AACR2

333,8
P769

To my mother and father

Published in 1982 by Praeger Publishers
CBS Educational and Professional Publishing
a Division of CBS
521 Fifth Avenue, New York, New York 10175, U.S.A.

© 1982 by Praeger Publishers

23456789   052   987654321

Printed in the United States of America

83-6139

Foreword

After a decade of confusion in energy policy, the United States is beginning to move toward a more balanced energy strategy. This welcomed trend is the product of careful thinking about the nature of the energy problem we face. Such thinking is to be encouraged.

Energy policy results from the policy makers' perception of the energy problem. If the problem is perceived to be the level of oil imports, then the answer is anything that reduces oil imports. The energy independence programs of Presidents Nixon, Ford, and Carter spawned hundreds of expensive programs--the synthetic fuels program, for example --that trampled considerations of environmental risks. On the other hand, if the problem is defined as vulnerability to disruptions in our supply of oil, then the answer is oil storage. Rapid fill of our strategic petroleum reserve is an example of a vulnerability-reducing measure. Careful thinking about the nature of the energy problem can lead to different perceptions of the problem and thus to quite dissimilar policy implications.

If those of us who subscribe to the vulnerability-reducing thesis are correct, then the country need not rush headlong to develop every available energy source regardless of the financial or environmental cost. Subject to certain constraints, we can allow healthy competition among the many sources of energy to guide energy policy. As long as the government's environmental and competitive guidelines are enforced, informed energy consumers can be free to choose from the available mix of energy sources according to their own preferences and circumstances. Similarly, as long as the government's environmental and antitrust guidelines are followed, energy suppliers should be free to develop any energy resource they think they can sell.

For oil and gas exploration, this energy policy has clear implications. First, there is no overriding national need to cast caution to the wind; energy producers should develop offshore oil and gas supplies at a rate dictated by economic efficiency and environmental protection. Second, the clear role for the state and federal governments is to set environmental guidelines and provide a climate that fosters responsible development of state and federal resources. These

vii

guidelines should be determined by state and federal officials based on costs and benefits accruing to the state and nation. The high price of energy today provides enormous incentives for producers to explore for and produce oil and gas. Since the states may bear a portion of the expected costs of offshore exploration--roads, pipelines, and other support structures, for example--perhaps the states should also receive a portion of the benefits derived from exploration on federal lands off their coasts.

These are among the issues examined in this book. Drawing on environmentalists, industry representatives, and federal and state officials, Dr. Goldstein has provided us with a vitally important book with a wealth of ideas, facts, and questions pertinent to offshore oil and gas exploration. Just as careful thinking has helped identify appropriate federal policies in the energy area, informed thinking can only improve the decision-making process regarding offshore energy development. Those in responsible positions should find this book necessary reading for the understanding of national and regional problems, and for the subsequent drafting of rational policies. Dr. Goldstein has demonstrated these exemplary skills in her earlier work, *ENVIRONMENTAL DECISION MAKING IN RURAL LOCALES: The Pine Barrens.*

Senator Bill Bradley

Senate Committee on Energy
and Natural Resources
Washington, D.C. 20510

Acknowledgments

There were many who contributed to the completion of this book. In the final stages of the manuscript, I was given support and encouragement from many of my colleagues at the Graduate Center, CUNY, and from my former Rutgers colleague, Ross K. Baker.

I wish to thank the U.S. Department of the Interior, the Bureau of Land Management for their help and cooperation. I appreciate the assistance of Richard Barnett and Judy Gresham in the New York OCS office, as well as the staffs of the California and Alaska OCS offices (all of whom supplied necessary maps and documents), as well as Peter Lafen of Friends of the Earth.

The acknowledgments would not be complete without noting the fine work of my publisher, Praeger, and the contribution of intelligence and humor by my editor, John Lambert, and the quality work of the production editors, Gordon Powell and Linda Berkowitz. Finally, I wish to thank my good friend, Dr. Timothy Johnson, Congresspersons Millicent Fenwick and Jim Florio, and Governor Thomas Kean, for their support of my work on the OCS Technical Advisory Committee on offshore oil leasing.

Contents

Introduction

The Gods are Just. No doubt. But their code of law
is dictated . . . by the people who organize society.
                    Aldous Huxley, *Brave New World*

It was prophetic that the state of California, the golden
shore, was the site of the first offshore oil well, drilled
off Summerland in the 1890s.

Gazing at this curious structure from beneath their top
hats and parasols, turn-of-the-century Californians viewed
the rig poised in the nearby shallow waters more as an at-
traction than as the forerunner of major technological and
social change it was to be late in the upcoming twentieth
century. But it was not until late into the century, in the
1970s, that the U.S. energy policy was even recognized as
central to the performance of the domestic and global political economy; and it was not until the latter half of the
1970s that a national energy policy was developed formerly
outside the hegemony of the private sector, that is, outside
the purview of the multinational corporations. That policy
centered on the activities of offshore oil and gas leasing
and exploration.

There have, of course, been energy policies formulated
all during the century: coal, a highly labor-intensive industry, has been regulated by the federal government along
questions of labor relations and conditions; and natural gas
has been regulated stringently in terms of interstate distribution and commerce. But there has been a critical absence of a unified national energy policy that would deal
either with production, regulation, the environment, or with
the needs, costs, and benefits to the U.S. society at large.

No more dramatically was this gap in energy policy felt
than in the winter of 1973-74. After a failure to achieve a
victory in one of the numerous wars and skirmishes with the
State of Israel, the Arab nations involved in the oil car-
tels, OPEC (Organization of Petroleum Exporting Countries),
and more particularly, OAPEC (Organization of Arab Petroleum
Exporting Countries), agreed to an oil embargo against those
countries not effecting a pro-Arab policy. These countries
included the United States, then importing 36 percent of
their crude reserves from OPEC, and several western European
countries, the Netherlands in particular.

xiii

The events that followed this oil embargo, namely the social and economic disorganization resulting from the sudden
shortage of supplies, led to a number of unintended consequences. For example, once the six-month embargo had ended,
OPEC moved to raise dramatically the price of their commodity. By January 1, 1974, the price of a barrel of crude oil
at the Persian Gulf had more than doubled since the winter
of 1973, and by 1975 it had quadrupled. Moreover, in the
economic sphere, after more than a decade of stability, the
cost of living in the United States began to rise precipi-
tously: whereas the rate of increase had been 4.1 percent in
1971, and 3.2 percent in 1972, in 1973 it rose to 6.2 percent. Then, in 1974, the cost spiral accelerated sharply.
That year saw double-digit inflation for the first time in
many decades, with the cost of living increasing at a rate
of 10.9 percent (Caplovitz, 1979).

In Caplovitz's study of U.S. families coping with the
stagflation of the 1970s, he makes the following observation
concerning citizens' growing distrust in government's ability to resolve the problems: "Many people blamed the federal
government for the economic ills of inflation and recession
and therefore it is not surprising that those who were hard-
est hit by the inflation had the least confidence in government." (Caplovitz, 1979)

It seemed perplexing that first a shortage and later a
price hike in only one-third of U.S. crude oil supplies
could ignite such a series of rapid changes and consequences. Debates in Congress and throughout the nation began to question the real nature of the energy crisis. Did
the political actions of OPEC merely serve to uncover a far
greater concern: were we in fact running out of oil and natural gas? Or was this the machinations of oil companies to
manipulate the market forces in favor of deregulated pric-
ing? In the present 1982 oil "glut" we raise those same
questions.

The speculations by former Secretary of the Interior
Stuart Udall during a series of policy roundtables conducted
by the American Enterprise Institute for Policy Research in
1975 implied that the U.S. oil industry could not fulfill
the energy demands of the country, even if they were given
complete price decontrols--decontrols for which the industry
was lobbying at that time. Thus, Udall concludes,

. . . even if we gave the [oil] industry complete de-
control, the industry still could not maintain our reserves, and our present production. . . . We will
still have to conserve. . . . industry people are in-

correct to say, "If you just rip off the lid and let us have these high prices, we will produce more and the crisis will be over because we will increase production." It will not happen.

The oil industry invested $4 billion more in 1974 than in 1973. Drilling increased 23 percent in 1974. But in 1974 we used 3.5 billion barrels of oil while the industry found only 2 billion barrels of oil. [Roundtable, 1975, p. 12]

While Udall and others questioned the resource potential of the U.S. industry, industry representatives lay the blame at the feet of unstable foreign producers. Thus, Robert Dunlop of Sun Oil argued Udall's point,

. . . since World War II, there have been eleven interruptions of international petroleum movements—a clear indication that it is essential for the U.S. to avoid becoming too dependent on overseas oil. The only real assurance of continuous supply of petroleum to the U.S. is to maintain an active, viable North American petroleum industry. . . . [Roundtable, 1975, p. 47]

While the oil industry pushed for price decontrols (which they were to gain in 1981 under President Reagan), the states themselves were deteriorating under the combined pressures of inflation-recession (stagflation) and the accelerating cost of energy. At that same forum in 1975, Senator Brook of Massachusetts noted that unemployment in his state was 13.9 percent, and that the cost of energy was so high in New England that it restricted industrial development.

Yet despite these critical conditions, a federal energy policy had still to be formulated. The enemies, it was popularly thought, were the Arab nations in OPEC, or, in some cases, depending upon the bias of the observer, the Israelis. Still others blamed a U.S. oil industry that had profited considerably from the shift in resource availability.

U.S. citizens began to respond to the social and economic stress brought about by the rise in gasoline and heating oil. Superpatriotism was reflected in the underlying hostility. The citizens were bewildered by the course of events and even identified environmentalists as the enemy because the group resisted the expansion of offshore oil exploration. For example, testifying at a hearing on environmental impact assessment in 1976, one man stated that he would sacrifice the entire coastal strip in the name of Uncle Sam. "I own beach land, sure, it could be contaminated, but what do I care. I want Uncle Sam to be good and strong and free; and if a little contamination is going to harm this country, well, that's just too bad. We can clean it up."

In summary, then, the shift in the Arab OPEC policies in terms of the politics of oil distribution led to dislocation of the economy and the society. But most importantly, this crisis of the mid-1970s led to a series of questions moving toward the need for a comprehensive U.S. energy policy. The efforts at drafting such a policy were fragmented at best.

As an example of the fragmentation, President Nixon offered the nation Project Independence, a concept that promised full U.S. energy independence by 1980. It was a goal that, in 1982, seems a remote possibility. And later, his successor, President Ford, presented a five-year program with essentially the same goal. Moreover, Nixon had established the Federal Energy Agency (FEA), whose role it was to oversee the oil industry in terms of domestic production, refining quotas, allocations and prices. However, at the same time, Nixon vetoed the windfall profits tax and rolled back domestic prices on "new" oil, that is, oil that had been drilled after 1976. These policies tended to continue the status quo and therefore did not alleviate any of the problems.

Before leaving office, President Ford presented to Congress the bare skeleton of an energy plan that would include, among other goals, the extended development of offshore oil and gas exploration. By the time Carter entered the presidency, he was aware that, one way or the other, a U.S. energy policy was required of his administration. Accordingly, Carter proceeded to elevate the status of the Federal Energy Agency to that of a department, and to pursue among a potpourri of solar, nuclear, and conservation programs, a special five-year plan for the leasing of offshore tracts to the oil companies for the purposes of locating and recovering petroleum basins beneath the continental shelves off the Atlantic, central Pacific, and Alaska coasts. These areas are referred to as frontier or wildcat areas in oil jargon since they have never before been explored for resources. What made this program especially unique was that it included, for the first time, participants in the planning process that went beyond the traditional purview of the private sector: the industry, and the federal government. Those new participants included environmental organizations and key representatives of affected states.

It was not accidental that such changes in the composition of decision makers came about. The decade of the 1970s
had been an era of legislative policy changes, and these
changes had focused on the impact of industrial development
on the natural environment. The period from 1974 to 1981,
in fact, was a critical period in national ocean policy.

The Coastal Zone Management Act of 1972, for example, administered by the Office of Coastal Zone Management, National Oceanic and Atmospheric Administration (NOAA) of the
Department of Commerce, provided grants-in-aid to states for
the development and implementation of management programs to
control land and water uses in the coastal zone. The policy
that the act established was aimed at balancing protection
of the coastal environment with the development and economic
interests. Moreover, the OCS (outer continental shelf)
Lands Act as amended in 1978, required a balance between oil
and gas development, and other ocean uses. It also called
for the extended participation of affected coastal states
and environmental groups, albeit in an advisory capacity.

This book is a study of that particular OCS national energy policy, including its evolvement in the Carter years
and changes now taking place under President Reagan. As
such, it is a presentation by all the levels of complex organizations involved in the drafting and implementation of
the OCS Oil and Gas Leasing Program.

We have included all the actors in this drama: environmentalists; state representatives; federal agencies, including the U.S. Fish and Wildlife Service, and the Bureau of
Land Management; and, of course, the petroleum industry. It
is a series of essays written by active OCS participants,
practicioners and experts, each of whom represent a different level of government, both state and federal; environmental organization, or the industry. As such, all the actors
in the process are also the writers of this series of
papers.

The book is divided into four parts, designating four
distinct perspectives. Part I, the environmentalists' perspective, offers an essay by Liz Kaplan, legislative director of Friends of the Earth, and examines in depth the fight
over renewed drilling off the California coast. Strictly
speaking, California is not considered a frontier area since
leasing for offshore oil and gas rights had been in process
for some 30 years in the Gulf of Mexico and off southern
California. But the tragic accident of the Santa Barbara
oil spill in 1969 left the citizens of California wary of

any further exploration. This incident in Santa Barbara was major in its proportions and influenced the decision to reduce activities of offshore explorations in the succeeding years. So major was this incident that it is duly noted in every section of this book by each author in turn.

With the spectre of reopening the drilling off the California coast in the new OCS program, activism and conflict ensued. The history of this struggle is contained in the Kaplan piece.

Part II, the states' perspective, offers two essays. The first essay in this section is a detailed history of the intense legislative battle over leasing rights in Georges Bank, New England's outer continental shelf region. It is written by Charles Colgan, economist-planner for the state of Maine, and offers an excellent perspective on the struggle between levels of governmental jurisdiction and industries over control of the planning process and ultimately, the resource.

The second essay in this section is written by Edward Wilson, of Virginia's OCS office. He describes the seldom reported tale of the Middle Atlantic Governors Coastal Resources Council (MAGCRC), a secret organization of governors in the mid-Atlantic region who organized and generated influence upon federal policy makers in terms of representation in decision making. The chapter provides an illuminating portrait of the governors of the Atlantic states in the process of organizing themselves to force the issue of representative power on the federal planning level.

Part III presents the federal perspective. Chapter 4 covers a debate on the Alaska issues as seen through the eyes of Esther Wunnicke, the states' manager of the federal OCS program. Specifically, the chapter notes the consideration given the host communities in that sparsely populated portion of the country, in terms of impact from development. Chapter 5 is by Frank Basile and Barbara Karlin. Basile is the Bureau of Land Management's Manager of the New York OCS office, and, at this time, overseer of the total Atlantic Coastal Program for the Department of the Interior. Ms. Karlin is the public information officer. These two discuss environmental issues and look at two negative themes pervading the reactions of the public to oil drilling and development off the mid-Atlantic coast, and they offer some findings since the mid-1970s.

Chapter 6, written by Bert Brun, U.S. Fish and Wildlife Service concludes the federal perspective. Brun's chapter centers on the mid-Atlantic region. He offers considerable detail and description of the potential environmental problems associated with the effect an oil spill would have on the life beneath the sea, and indeed much of this discussion is applicable to other coastal regions.

The fourth part is, appropriately, the industry perspective, authored by Shell Oil Company's O.J. Shirley. Shirley is manager of government and industry relations, and includes the writings of several key oil company representatives in his argument that favors offshore oil drilling. It is here that we can view the fourth side of the prism, the industry perspective, a position frequently absent from studies on energy policy. In this segment, the author attempts to respond to the issues raised in the preceding chapters concerning regulation, production, and the environment.

The final chapter, "Landlords of the Sea," is written by this editor, Joan Goldstein, sociologist and appointed member of the mid-Atlantic OCS Technical Working Group. This summary section pulls together the consistent thread in the drama that moves throughout the book: namely, the social, economic, and political forces that influence the decision-making process. We may ask, and rightly so: who gets involved, who benefits, gains, or loses in the types of political arrangements achieved? How does this evolving web of relationships influence the interconnections between complex organizations, that is, between competing industries and competing levels of governmental jurisdiction? The central question is, after all, can we sustain multiple uses of the sea? Can the fishing industry and the oil and gas industry mine the same resource without destroying the resource or each other? This last question is vital to the writers in this book and to the population affected by new energy policies. For it is this policy, multiple uses of the sea, that is fostered by the present Reagan administration through the actions of the Secretary of the Interior, James Watt. In recent months, Watt has streamlined the OCS activities to accelerate drilling activities and to shorten the process of environmental-impact assessment.

Since the process in study is actively in a state of change, this book offers a unique opportunity for the reader to gain an understanding of recent energy policies with respect to offshore exploration activities. Moreover, it offers the reader a leg-up on the state of the art of energy policy planning. In this sense, it may be possible to understand the process and evaluate the results.

REFERENCES

Books:

Caplovitz, David. *Making Ends Meet: How Families Cope with Inflation and Recession.* Beverly Hills, Cal.: Sage, 1979.
Engler, Robert, ed. *America's Energy.* N.Y.: Pantheon Books, 1980.
Huxley, Aldous. *Brave New World.* N.Y.: Harper & Row, 1969.
Steinhart, John and Carol Steinhart. *Energy Sources, Use, and Role in Human Affairs.* North Scituate, Mass.: Duxbury Press, 1974.
Szuc, Ted. *The Energy Crisis.* N.Y.: Franklin Watt Inc., 1974.

Documents:

Bureau of Land Management. Environmental Impact Statement Hearing. New York, Jan. 29, 1976.
Energy Policy Roundtable, American Enterprise Institute for Public Policy Research, Washington, D.C. Oct. 2, 1975.

Newspapers:

New York *Times.* 1981. October 15.
New York *Times.* 1981. February 22.

PART I

THE ENVIRONMENTALIST PERSPECTIVE

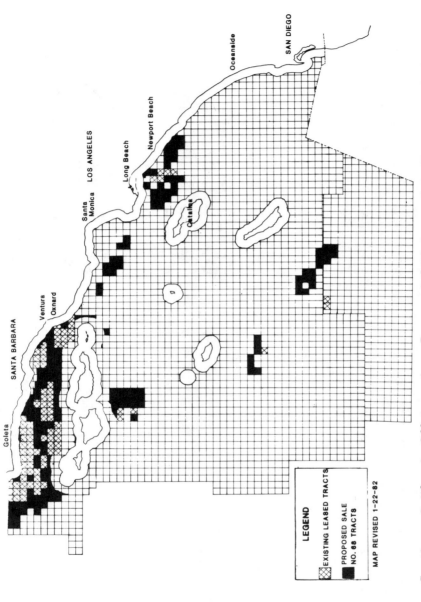

Southern California Offshore Area Proposed OCS Lease Sale Number 68 (12/23/81)

LEGEND

EXISTING LEASED TRACTS

PROPOSED SALE
NO. 68 TRACTS

MAP REVISED 1-22-82

SANTA BARBARA

Goleta

Ventura
Oxnard

Santa
Monica   LOS ANGELES

Long Beach

Newport Beach

Catalina

Oceanside

SAN DIEGO

1
California: Threatening the Golden Shore

Elizabeth R. Kaplan

It is little wonder that California is the most progressive state in addressing the fate of its coast. From the magnificent bluffs of Point Reyes National Seashore to the coast and islands of the Santa Barbara Channel, home to thousands of marine mammals and millions of birds, the California coast is awesome in its beauty. Californians more than any other Americans define the quality of life in terms of easy compatibility with the natural world. The California coast is lucky, for its inhabitants and neighbors are a highly-politicized population, the first to see that a problem exists, the first to wrestle with it and forge innovative solutions. Other states will follow; California will lead, is the widely held notion, and not without reason.

The short history of offshore energy exploration in California coastal waters could hardly contrast more with that of the Gulf of Mexico. The federal offshore program in the gulf grew slowly from the 1940s in quiet obscurity, unexamined by the public eye, and only cursorily monitored by the federal government, until the proposal to designate the unique Flower Gardens coral reef a National Marine Sanctuary right in the middle of a drilling area off Louisiana and Texas catapulted the federal offshore program into the national press. But that is another story.

The federal offshore program had barely begun in California when a well in the Santa Barbara Channel blew and 3 million gallons of oil washed across miles of beautiful beaches and marshes killing at least 1,500 birds and threatening a valuable fishery and recreation area. Santa Barbara was a town with a proud history of tight zoning restrictions over urban sprawl, a town that treasured its lovely vistas of the Pacific Ocean and was determined to protect them.

When the Department of the Interior began leasing for oil and gas in the channel in 1967, this new threat gave rise to considerable alarm in the community. Santa Barbara already had a 16-mile-long, 3-mile-wide state marine sanctuary along the coast as protection against state drilling in the tidelands. In 1967 county officials approached the Department of the Interior and asked for a one-year leasing moratorium. Objections to further leasing were grounded primarily on aesthetics. In 1967 environmental concerns about the risks of offshore drilling had little scientific research to look to for answers. It was known, however, that several faults ran through the channel seabed, which raised local concern about the safety of drilling there. Furthermore, oil spill containment and cleanup technology was still primitive, and the abundant populations of birds, marine mammals, and fish could be at considerable risk. Most clearly articulated, however, was the Santa Barbarans' antipathy to ugly drill rigs marring their ocean vistas (Baldwin and Page, 1970).

In the fall of 1967 the Department of the Interior bowed to local concerns and announced that it would prohibit leasing in a two-mile buffer zone next to the state sanctuary. The county requested further delay and other restrictions, but the government proceeeded to offer one-half million acres of channel lands in December.

By March, 1968, the Union Oil Company had discovered oil and shortly obtained permits from the Corps for construction of the infamous Platform A. On January 28, 1969, as the drill for the fourth well was being withdrawn, gas and oil gushed up the well, and the well went out of control.

The Department of the Interior moved swiftly to impose new safeguards on drilling and created a moratorium on drilling in the channel. Several weeks later Secretary Hickel established a 21,000-acre Ecological Preserve and a 34,000-acre buffer zone in the channel next to the state sanctuary. In the meantime, dead and dying birds were being plucked from the oily water at an alarming rate by an army of volunteers.

In April drilling resumed in some channel areas, but seepage from the well area continued through the fall. Estimates of total gallons spilled has varied, but one expert estimated the spill at 3,250,000 gallons. This was small compared to some spills such as the Torrey Canyon or the infamous Ixtoc Gulf spill, but its political implications were enormous. Studies of the spill have concluded there were negligible long-term effects, but it has also been pointed out that the lack of pre-spill base-line data makes it extremely difficult to assess the damage accurately. Furthermore, scientists admit they do not know the long-term effects of oil in bottom sediments or its long-term impacts on aquatic food chains.

Santa Barbara has never been the same, however. Within hours of the blowout, Get Oil Out (GOO) a citizen's based organization was born. Still lively a dozen years later, GOO has had an impressive history of litigation and public participation in the leasing process. In 1973, GOO sued the Department of the Interior to require site specific environmental impact statements in the channel leasing. Eventually GOO lost the case, but the Department of the Interior has done them anyway. In 1971 GOO helped write the state law that created a three-mile state sanctuary around the Channel Islands, and they were successful in getting a provision into the Outer Continental Shelf Land Act (OCSLA) amendments of 1978 that required that drill rig workers must be certified.

The Santa Barbara area has been one of California's most heavily targeted areas for energy production and related facility projects. But this growth has not come without conflict. A 1968 oil-processing plant was doomed by a county wide referendum as well as a big residential development on El Capitan Ranch overlooking the Pacific in 1970. But in 1975 Exxon won the right to put an oil-processing plant in Las Flores Canyon after a bitter fight. When all else has failed, Santa Barbarans have been known to confront the enemy head-on. On Easter Sunday, 1969, after the big spill in the channel, a contingent of concerned citizens sat in front of huge oil supply trucks on Stearns Wharf to prevent them from reaching their destinations (Sollen, 1978).

The Santa Barbara area has also been chosen for a number of nuclear power project sites, and one of its hottest issues has been the attempt by the gas utilities to get a Liquified Natural Gas port at Point Conception. In the mid-1970s Lease Sales 35 and 48 proceeded apace, and oil platforms have continued to spring up in the channel, one of the most heavily traveled tanker lanes in the country. Development conflicts in Santa Barbara continue to burden a citizenry that ranks as one of the most actively involved in land-use decisions in a very active state.

What troubles Santa Barbarans, who feel that they are bearing an unfair share of the contribution toward the national goal of energy self-sufficiency, is that the channel produces about 12 million barrels of oil per year, while Americans are consuming nearly 20 million barrels of oil per day. In short, is there enough oil out there to make it worth the risks to the environment?

The Santa Barbara blowout had state-wide impacts as well. It galvanized local activists to work harder on achieving coastal protection. In 1972, through a state-wide referendum, California passed Proposition 20 creating a unique state-wide Coastal Commission with a well staffed central office and six regional commissions. The commissions were given authority over almost all types of development along a thin strip of the coastline and extending seaward three miles to the limits of the state jurisdiction. State agencies as well as private developers had to appeal to the commissions for permits for their activities.

The Coastal Commission's primary task was to draw up a plan outlining the acceptable uses along the entire California coast and submit it to the state legislature in 1976. Offshore oil development was only one of the many uses to be permitted in the coastal zone, but the formation of the commissions created a public climate in which organizing against offshore development was made easier. Interested citizens became familiar with the public processes of hearings, filing appeals of local commission decisions to the central commission, and lobbying members of the local government for various positions on permit applications. The commissions provided rallying points and an abundance of information to constituencies who before had rarely talked to one another: fishermen, homeowners, and environmentalists who all shared concerns about certain projects, often for different reasons. Access for instance, was as big an issue to local townspeople as protecting the environment, often bigger (Healy et al., 1978).

When the California Coastal bill finally passed the state legislature in 1976, it was not without a harrowing fight against a powerful lobby of development interests. It put in place a system of immediate citizen access to decisions pertaining to coastal uses, which became a framework for future OCS citizen activism.

California's outer continental shelf comprises only a small fraction of the roughly 1 billion acres of OCS that the U.S. government would like to see explored for potential oil and gas. By the end of 1980 California had produced only .2 billion barrels (bb) of oil and .1 trillion cubic feet (tcf) of gas compared to 5 bb of oil and 43.9 tcf of gas from the Gulf of Mexico. Total unrecovered resources from the California OCS are estimated at 3.7 bb of oil and 6.7 tcf of gas compared to 12.3 bb of oil and 64.4 tcf of gas from the Alaskan OCS and 6.6 bb of oil and 71.9 tcf of gas yet to be found in the Gulf (GAO-EMD-81-59). In another state, federal plans for similar size exploration would probably have stirred only minor protest. But the California OCS experience centers in the Santa Barbara area, where a catastrophe of awesome proportions in 1969 politicized the state and made it wary of future development.

NATIONAL REPERCUSSIONS

The Santa Barbara blowout had national repercussions for the federal OCS program. The moratorium delayed a number of lease sales in other parts of the country for a year. By 1969 southern California had only two OCS leases, which had been relatively uncontroversial. The Santa Barbara blowout changed all that. Instantly the people of Santa Barbara became politically active to fight the issue that represented careless destruction of their environment, and repercussions of (the blowout in Santa Barbara) have been felt in the passage of the National Environmental Policy Act (NEPA), the California Coast Act, Coastal Zone Management Act, and the OCS Land Act amendments in 1978.

The early 1970s experienced conflicting pressures on OCS development at the same time that environmental consciousness toward offshore drilling was rising. In 1974 the United States experienced the first oil embargo by the OPEC countries, sending the country and its policy makers into a panic over our energy future. One of the key elements of the new strategy for energy self-sufficiency was, of course, rapidly expanded OCS development that was to be conducted in a safe environmental system of checks and balances through amending the OCS Land Act of 1953. After four years of national debate the OCSLA amendments were passed in 1978 creating a careful process of environmental impact analysis at every step of the OCS development process.

The OCSLA amendments called for a considerably expanded OCS program and directed the president to create a five-year national OCS plan. At the same time they included fairly stringent environmental safeguards and guaranteed that the public and state and local governments could participate in the process. It did not, however, give the states authority to override the federal government in the selection of sites for leasing. It merely gave the states ample opportunity to comment on every step from tract selection to exploration and development, as well as transportation and support-facility development. Between the lines was the unspoken yet historic tradition of federal government representatives honoring state concerns in federally-assisted, local-development programs. After all, the history of OCS development had been almost free of controversy until the Santa Barbara experience, and with careful environmental analysis of each step in the process, it could once more become free of conflict, or so it was believed.

During the 1970s the national leasing program became increasingly ambitious as the energy crisis developed. At least five different leasing goals were promulgated and nine different schedules published over the decade, each aspiring to greater production. These many schedules illustrated a frontier program groping for a systematic buildup of production and continuity. Under the Carter administration the first schedule was developed in which state governments were given a 60-day comment period prior to final issuance of a sale.

The 1979 and 1980 five-year plans incorporated the requirements of the OCSLA amendments and were given wide public review. Only 13 percent of OCS development in the 1980s was slated for the California coast, while emphasis was placed on frontier areas, especially Alaska, the North Atlantic, and mid-Atlantic, which had had no previous OCS exploration.

After the moratorium, the initial reopening sale in southern California was Sale 35, planned for September 1974. But the state sued maintaining that the requirements of NEPA had not been met. In late 1975, the court ruled against the state. But in the meantime two proposed southern California sales, the 1976 and 1978 sales were dropped. A second southern California sale, Sale 48, stirred considerable controversy and was delayed two years until 1979, after a suit by the county of San Diego failed to stop it.

MARINE SANCTUARIES BRING PROTECTION AND CONFLICT

Among the visionary environmental laws passed by the Congress was the Marine Protection, Research and Sanctuaries Act of 1972. This law established a Marine Sanctuaries program to be administered by the National Oceanic and Atmospheric Administration (NOAA) in the Department of Commerce. It was dedicated to protecting marine areas of special values and unique qualities. A site could be chosen for biological, research, recreational, or simply aesthetic reasons. In establishing a sanctuary the president was given the authority to provide any appropriate regulations for the area, including prohibitions of oil and gas development. In fact, in passing the law, Congress specifically sited the increasing threat of oil and gas development as a major reason for needing a sanctuary program.

The Marine Sanctuaries program languished in virtual disuse until the Carter administration. Then a half dozen sites were given serious consideration, three of these were in California. They were the waters around the Santa Barbara Channel Islands, the waters around the Farallon Islands near San Francisco, and the Monterey Bay area.

Blessed with a temperate climate, California historically boasted an incredible variety of coastal birds and other wildlife. But the increased population density of California destroyed much habitat and inadvertently created two unique and natural wildlife refuges. As population areas grew, the coastal animal populations that were able to, fled the coast for the safer havens of the offshore islands. Thus, both the Santa Barbara Islands and the Farallon Islands developed extraordinarily rich populations of birds, marine mammals, and other wildlife. While most of these islands had been bought by the state, the federal government, or the Nature Conservancy in recent decades, increased recreational activity in the area was gradually bringing dangerously high levels of human pressures into the island waters.

It was no coincidence that the County of Santa Barbara sent to NOAA in 1978 a detailed nomination for a Marine Sanctuary off the Santa Barbara Islands including the Santa Barbara Channel at the same time that an Environmental Impact Statement (EIS) was being prepared for Lease Sale 48 in those waters. This was not a case of federal intervention in state affairs but rather a regional governing body begging the feds to help protect its incredibly valuable resource in the face of intolerable development pressures. No fewer than nine major energy related projects were being planned in the Santa Barbara area as well as increased OCS activity. Despite the fact that the Santa Barbara Islands themselves were protected, the threats to the waters around them was increasing rapidly. Unique marine mammal and bird populations were being seriously threatened.

San Miguel Island supports the world's largest and most diverse pinniped community. California sea lions, harbor seals, northern fur seals, northern elephant seals, and Stellar sea lions all breed on the islands, and the endangered Guadalupe fur seal has been known to haul out at Point Bennett. Nine species of marine birds breed in colonies on several of the islands. Over 168 species of marine birds inhabit the channel and coast. The Santa Barbara Island

chain is also a major stopping place for migratory birds.
Seven rare and endangered birds rely on the channel's habitats, including the least tern and two varieties of rails.

In addition to seals and sea lions, most of the endangered whales of the world either live or migrate through the
channel area. Gray, sei, sperm, blue, fin, humpback, and
Pacific right whales have all been seen, and some are regular inhabitants (Santa Barbara County, 1978).

The threats to the channel and its diverse animal life by
OCS development were amply described in the Draft Environmental Impact Statement (DEIS) of Lease Sale 48. "According
to the oil spill risk model for this lease sale, a major
(1,000 barrels or more) oil spill anywhere within the channel area would hit San Miguel Island" (California Comments
on Lease Sale 53, 1980).

The Bureau of Land Management (BLM) oil spill model predicted for Lease Sale 48 3.1 spills of more than 1,000 barrels and 6.05 spills of 50 to 1,000 barrels. Santa Rosa and
Santa Cruz Islands have a "near certain probability of being
oiled from leasing in the Santa Barbara Channel" (California
Comments and Findings on Lease Sale 53, 1981). For the
first time, oil exploration leases would be sold near the
northernmost islands, the most heavily inhabitated by birds
and marine mammals. BLM noted that this could cause serious
disturbances to several species leading to reduced reproduction or even elimination of species.

Yet to be mentioned are two other economic interests in
the Santa Barbara area, the recreational community and the
fishing community. The islands themselves are of considerable recreational value, and scuba diving and recreational
fishing are large commercial activities in the waters surrounding them. In addition, 6,000 tons of fish were caught
commercially off the northern islands in 1975. According to
state figures some 450,000 fish were caught by sport fishermen in 1974. These numbers have surely increased since
then.

While the oil companies went to great lengths to convince
the doubtful that the Santa Barbara 1969 spill had caused no
lasting damage, the many members of the fishing community
believed otherwise. Their greatest worry, however, is evidence of chronic small leaks and spills from the platforms
that create regular small slicks and deposit oil in the
sediment. The industry claims that natural seepage is the
culprit, but proximity of oil to platforms has convinced the
fishermen otherwise. Repeatedly, trawler nets have come up
covered with oil, and oil infested dead crabs have been
found floating near platforms. In places the ocean bottom

looks like it has been stripmined, according to fishermen, where the dumping of drill muds has suffocated all life on the bottom. Equally disenchanting are regular entanglements with the flotsam and jetsam of the rigs--old tires, cables, refuse, and capped well heads.

The Santa Barbara marine sanctuary proposal achieved enormous local support, including all the local governments, the city and county of Santa Barbara and several local congressmen plus Senator Cranston. Surprisingly, only the fishing community resisted the idea, fearing that the sanctuary regulations would create additional restrictions on local fishing operations. In the end, however, when NOAA made clear that it did not intend to regulate fishing in any way, the fishing community gave tacit if not outright support.

The proposal for a marine sanctuary for the Farallon Islands proceeded on a similar course, while the Monterey Bay proposal, which did not achieve such near unanimous local support, was temporarily left behind. Like the Santa Barbara Islands, the Farallons are habitat extraordinaire for nesting sea birds and several varieties of seal and sea lion. Virtually the world's entire population of ashy storm petrel, a magnificent rare sea bird, nests in the islands. Probably over half of California's nesting seabirds nest in the Point Reyes-Farallon Islands area. Some 23 species of marine mammals have been sited in the area, and its bird population contains some of the largest sea-bird rookeries in the continental United States. The world's entire population of gray whales migrates through the Farallon Islands Channel twice each year. A sanctuary here was viewed as a perfect complement to the Point Reyes National Seashore and Farallon Islands National Wildlife Refuge.

Both the proposals that moved forward contained fairly detailed management plans, which included management by the state of California and a clear exemption from any additional fisheries management. The key component in both was a prohibition on oil and gas drilling within the sanctuary boundaries. When NOAA sent its proposals to the Office of Management and Budget for interagency review, the Department of the Interior objected strenuously to these prohibitions on the grounds that they set a bad precedent and conflicted with the national goal of speeding our national energy self-sufficiency. NOAA argued that there were no leases either within the boundaries or contemplated there because the secretary had already deleted from Lease Sale 48 all tracts which lay within six nautical miles of the Santa Barbara Islands, and he also deleted all tracts from Lease Sale 53

within the Point Reyes-Farallon Islands Sanctuary area at
the request of state and local governments. Because of the
interagency disagreement, the two proposals were never
cleared to go to President Carter's desk, and a direct
appeal was made by the state to the president on both sanctuaries. In late September 1981, the president signed the
Santa Barbara Sanctuary designation complete with oil and
gas prohibition. The sanctuary boundary was a compromise
that included six nautical miles around the islands. The
original proposal had been for the entire channel, with all
state and local governments urging full channel protection.
But NOAA, which had no management capability, feared taking
on the enormous burden of regulating tanker traffic in the
channel and all the political pressures that would entail,
and backed off to a six-mile boundary as a gesture to both
the oil industry and The Department of the Interior.

The Department of the Interior Secretary Andrus carried
his appeal to the president personally to dissuade him from
going forward with the two sanctuaries if they would prohibit energy development, and for a time it looked as though
President Carter might let the clock run out on the Point
Reyes-Farallon Islands sanctuary. But a powerful appeal by
29 members of the California Congressional delegation, thousands of letters to the President, plus a personal appeal by
Senator Cranston moved the president to sign the sanctuary
designation in the final hours of his administration.

How much oil and gas was sacrificed for these sanctuaries? In the Point Reyes area, part of only two tracts were
deleted from Sale 53, which covers over 1 million acres.
Another 50 tracts considered to be of low potential that
might be leased in future sales exist within the sanctuary
boundary. While the Santa Barbara sanctuary meant that some
24 tracts would not be leased, the amount of recoverable
energy left in the seabed was small in relation to the entire projected production for Lease Sale 48, only about one-
fourth of a day of U.S. energy consumption. Clearly the
issue was not the oil lost, but the precedent-setting act of
putting any portion of the seabed off limits to energy exploration.

A number of things are instructive concerning the California marine sanctuaries. The Santa Barbara case marks the
unusual situation where not only the state but local governments sought a more conservation-oriented policy than the
federal government. Nearly 70 percent of the comments on
the Santa Barbara DEIS were in favor of prohibiting energy
development in the entire channel, but NOAA bowed to industry pressure and limited it to six miles, approximately one-

half to one-third the original size requested. The muscle of the oil and gas industry was felt heavily in Washington, but had little impact at the state and local level. Nor could big oil influence local representatives who overwhelmingly supported the larger sanctuary designation.

The California experience is in marked contrast to both the Beaufort Sea Marine Sanctuary proposal off the north slope of Alaska, and the Georges Bank proposal. These were developed by conservation and fishing communities and never received support from either state or, with few exceptions, local governments. Both of these sanctuary nominations were designed to limit and partly forbid oil and gas development in two incredibly productive biological areas, but without state support they never got off the ground. Industry pressure was too hard to withstand without strong backup by the states affected.

The success of the California designations, the first of national significance since the Marine Sanctuaries Program was created in 1972, was short-lived, as we shall see. The new Reagan administration moved swiftly to undo the work of the Carter administration, by suspending the oil and gas prohibitions, first temporarily and then indefinitely. The administration that campaigned on a states rights platform showed less interest in state concerns about OCS development than any previous administration, Democratic or Republican.

CALIFORNIA DISCOVERS CONSISTENCY

When legislation is shaped in the forge of Congress, it often happens that the most heated struggles are over amendments that contribute little to the future turnings of public policy and decision making. Likewise, sometimes an obscure clause virtually unnoticed by the Congress is the crucible from which national direction is determined.

So it was with the consistency clause of the Coastal Zone Management Act. In the heady days of the eary 1970's, protection of coastal and ocean resources received considerable attention in Congress, particularly in the wake of the Santa Barbara oil spill. But the pressures for a national interest in coastal protection policy ran head-on into the powerful states-rights representatives on Capitol Hill, who were not about to give away control of the coast to some eastern elitists in Washington. Thus a voluntary coastal management program was created in 1972 through the Coastal Zone Management Act (CZMA). States would receive federal money to plan for future development and protection in their coastal zones, but they would have to devise a plan that would meet federal standards to remain eligible for future grants. To nail home the states' autonomy in this program, a section was added that said that any federal activity that directly affects the coastal zone had to be consistent with the state's coastal management program. The federal agency had to certify that its activity was consistent with the state's coastal plan and submit it to the state for verification.

This little clause quickly became the center of a major power struggle between the states and the Department of the Interior, with the Departments of Commerce and Justice siding with the states. The consistency clause was totally appropriate to the philosophy of the law. It made little sense to give the states federal dollars to plan for the wise management of coastal resources if the federal government could come in and fund a huge sewage treatment plant on the edge of a productive estuary that the state had set aside for no development.

With the approach of Lease Sale 48, the next major lease sale in southern California, a critical question arose. Did preleasing activities related to offshore drilling come within the consistency clause of the CZMA? Or did the consistency clause apply only to the later exploration and development activities? Under California's state Coastal Act and the federal CZMA, the California Coastal Commission had authority to approve all OCS plans that affected the state coastal zone for consistency with their coastal plan. Because most of the Lease Sale 48 area already included the infrastructure to service increased OCS development, the state did not have major problems with the sale. Already in place were service bases, processing facilities, pipecoating yards and harbor facilities. The Coastal Commission approved all the plans in Lease Sale 48 except for one Chevron exploration plan because the tract was within 5.7 miles of Anacapa Island in the Santa Barbara Channel, the only breeding ground in the state for the endangered brown pelican. Chevron ultimately withdrew its plans to explore the tract, but went forward in the field elsewhere. The state's consistency review did cause several companies to modify their plans; in some cases they changed the location of proposed activities, increased on-site oil-spill containment or cleanup equipment or beefed up environmental assessments.

The commission could claim with confidence that it was not in the business of obstructing offshore oil development, but rather was a watchdog to make sure that all proposed activity was conducted in a manner that would not jeopardize California's abundantly rich, living coastal resources.

As a cap to its Lease Sale 48 position, the state in May 1979, asked the Department of the Interior to certify that the proposed Notice of Sale on 48 was consistent with the state Coastal Zone Management plan under Section 307(c)(1) of the CZMA. With this action the battle was joined, and a bitter struggle ensued between the state and federal government over ultimate control of OCS leasing. The Department of the Interior Secretary Andrus responded that the consistency clause does not apply to preleasing activities because such activities do not directly affect the coastal zone. Therefore no consistency determination was required.

California submitted that the actual notice of sale in fact did directly affect the coast because every other action flowed from the selection of tracts that were in the sale. Only by having some control over which tracts were actually leased could the state guarantee: a critical fish and wildlife habitat; that important recreation areas and hazardous tanker traffic areas be adequately protected from potential oil spills; and protection from other dangers and disturbances. Once the tracts were leased, an oil company had the right to develop them. Indeed, the due diligence clause of the OCSLA required that they be developed in a timely fashion. While a company's exploration and development plans could be submitted to the consistency test, the right to go forward on every tract in the sale had been gained. And if that right were taken away at some stage of the process, the company could claim considerable recompense from the federal government for its investment. From the state's point of view it made the most sense to eliminate tracts at the very beginning that might cause problems later on. California was fully aware that once an oil company had invested millions of dollars in a tract, the pressures against a negative consistency determination down the road would be enormous.

The Coastal Commission therefore developed criteria that it considered reasonable to meet the consistency requirements:

1. No tract could be within twelve miles of the range of the endangered sea otter.
2. There should be no leasing within six nautical miles of areas of particular significance to marine mammals or nesting marine birds in the Santa Barbara Islands area.
3. There should be no leasing within 500 yards of the coast guard established Vessel Traffic Separation Scheme. This would affect the heavily trafficked

Santa Barbara Channel, where platforms near the tanker traffic would increase the likelihood of collisions. [California Coastal Commission, Comments on Lease Sale 48, 1979]

The state negotiated successfully with the Department of the Interior on Lease Sale 48 to take care of its major concerns, and would have validated a consistency certification by the department, but the Secretary of the Interior viewed the consistency clause as a grave threat to its right to lease where it chose on the continental shelf. The Department of the Interior claimed that the OCSLA amendments provided ample guarantees of state participation and environmental protection during the preleasing stage. Consistency properly should apply only to the exploration and development stage because the sale of tracts had no physical impact on the coast. Therefore the sale did not "directly affect" the coast. Thus the Department of the Interior refused to make a consistency determination for Lease Sale 48.

In the meantime the Justice Department reviewed the controversy and concluded that the prelease Notice of Sale for 48 did indeed come under the consistency clause, but the Department of the Interior ignored that opinion. In June 1979 California appealed to the Department of Commerce to act as mediator in the dispute that the Department of the Interior agreed to. A public hearing was held in Los Angeles in September and California presented a detailed statement for the record. The Department of the Interior declined to come forward with an official position at that time. A month later the Department of the Interior produced its formal position just a few days before the formal mediation conference.

In February 1980, the Department of Commerce issued its opinion that the mediation had failed and that California was correct in requiring a consistency determination. But the Department of the Interior indicated that it simply would not compromise its position. Because no actual tracts were at issue in the Sale 48 dispute, the conflict simmered for a number of months until the next sale, Sale 53, came up for consideration. In May 1980, the California Commission once more wrote the secretary asking for a consistency determination for Lease Sale 53, scheduled for May 1981.

Lease Sale 53: Taking Off the Gloves

Lease Sale 53 was a brand new ball game for California. It was a frontier sale, which extended from the northern edge of the old sales in the Santa Barbara area all the way to the Oregon border, 700 miles of mostly undeveloped coastline. It covered some of California's most pristine and beautiful coast and it threatened to bring giant drilling platforms into one of the richest fisheries of the world as well as into recreation areas that attracted millions of visitors a year. It threatened the Point Reyes National Recreation area, the Elkhorn Slough Estuarine Sanctuary, numerous state wildlife refuges, and the three proposed national marine sanctuaries. Lease Sale 53 covered five resource basins encompassing 1.3 million acres, the largest being the Santa Maria Basin lying just north of the older lease areas from Point Conception north to Morro Bay. The other four were the Santa Cruz, Bodega, Point Arena, and Eel River basins. The northern end of the Santa Maria tracts ran right into the habitat of the endangered sea otter and therefore into a howl of public protest.

After the call for nominations in November, 1978, citizen groups and local governments up and down the coast began mobilizing to protest the sale. A group called the Coalition on Lease Sale 53 was formed to fight the sale; it represented 23 organizations, primarily environmental and fishing. Local governments as well as citizen groups were incensed. The County of Santa Barbara wrote, "In a short span of five years this frontier area has risen from being virtually unthought of (ranked 16th out of 17 areas considered) in 1974, to . . . seventh out of 22 areas." The reason for this initial low ranking was clear--low resource potential. The county pointed out that in several public statements Secretary Andrus had expressed doubts about leasing the area at all because of low potential and high environmental risks, and yet the proposed sale kept reappearing on redrawn plans.

Negative nominations, or tracts recommended for deletion from the proposed sale, were made by a number of state agencies as well as the Fish and Wildlife Service and National Marine Fisheries Service. But the people along the coast, whose concerns were more personal, sought outright abandonment of the sale. Local and national environmental organizations, fishing groups, recreation associations, businesses, and unaffiliated citizens turned out in droves to testify against the sale or wrote in comments opposing it.

An organization called the Commercial Fishermen of Santa Barbara, combat veterans of the channel leases, went to the state-wide Pacific Coast Federation of Fishermen's Association (PCFFA) and warned of their unhappy experience in the Santa Barbara Channel. The PCFFA represented 16 marketing associations and five fishermen's associations reflecting most of the state's salmon trollers, crabbers, herring gill netters, shrimp trawlers, and others. The California fishing industry contributes $600 million a year to the state economy and it was not difficult for the PCFFA to persuade the governor that Sale 53 would put that economy in serious jeopardy. The commercial fishermen were joined by sport fishing associations and other local economic interests. Even the tanker industry expressed its concerns about conflicts with pipelines and interference with traditional port access and the dangers of collisions. Ultimately, nearly 700 individuals commented on the proposed sale.

Lease Sale 53 brought together communities that are often at odds--environmentalists, commercial fishermen, tanker companies, sport fishermen, local and state government agencies--all united this time in opposition, in varying degrees, to the sale. It was a formidable frontal assault mobilized in a state that had considerable experience in organizing public pressure. The Congressional delegation was also quick to get involved. Congressmen Leon Panetta from Monterey and Leo Ryan of San Mateo became active early on against the proposed sale. Gradually, Panetta and Congressman John Burton developed a contingent of bipartisan Members from the California delegation that kept up a steady stream of pressure on the Department of the Interior.

It was difficult for the oil industry to accuse environmentalists of wanting to lock up an enormous national resource for the benefit of a few users. At issue were large economies that were seriously jeopardized by the arrival of giant drilling platforms up and down the coast and the concommitant development of necessary service infrastructures. The Secretary of the Interior pressed the urgency of the need to hasten toward energy self-sufficiency, hoping that the state and its citizens would rally to the cause. But up and down the coast of California, citizens and bureaucrats were doing their homework on the tradeoffs, which they presented in hearings and commentary, so that ultimately the federal case for a full-scale go-ahead looked foolishly weak.

Using both the Department of the Interior's draft Environmental Impact Statement and local research, various groups and state agencies presented projected oil-spill damage information concerning critical habitat areas that was quite devastating. If the entire Santa Maria Basin were leased for instance, it was virtually certain that there would be major damage to the threatened sea otter population. Other data was collected for critical wetlands areas, sea bird rookeries, and seal haul-out areas. The California

Air Resources Control Board found that there was no evidence that the oil companies could meet state air-quality standards in a number of areas. The State Department of Fish and Game noted that there still did not exist sufficiently sophisticated oil spill containment and cleanup technology to safeguard important habitat areas. Santa Barbara drilling was in relatively shallow, protected water compared to much of the Lease Sale 53 territory, and controlling a spill would be much more difficult.

The largely undeveloped Mendocino coast, the Farallons, and Santa Cruz coast were zoned primarily as agricultural land. The cumulative impacts of the lease sale had not been examined by the Department of the Interior, and the state saw its careful zoning plan for the coast about to be destroyed by the onslaught of industrial development. A decade after the Santa Barbara oil spill, a much more sophisticated methodology existed for projecting damage from OCS development. This time almost no one mentioned aesthetics. People's livelihoods were at stake.

At the same time the oil and gas industry was having difficulty projecting the actual potential of the area to be leased. No known reserves could be identified. Santa Barbara County did an analysis of Lease Sale 35 and other earlier leases in the Santa Barbara areas and showed that the original claims for recovery were way over the actual discoveries made. Gradually, agency by agency, hearing by hearing, the case was built that the risks to the state fishing, recreation, and other economies, not to mention the more difficult to define values of wildlife preservation, simply did not seem to be worth taking for such a modest amount of oil and gas to be recovered.

Serious economic questions were also raised concerning the saleability of offshore oil. All of the oil in Lease Sale 53 was expected to be high in sulfur. In 1979 California was producing a 500,000 bpd surplus of high sulfur oil with predictions that that number could double. The California Department of Energy predicted that it might be difficult to find markets for the high sulfur oil in an already glutted market. Furthermore, California's ability to refine high sulfur oil was severely limited. Thus, in all likelihood, this oil would be shipped to Japan where refineries for high sulfur oil were abundant. Risking the California coast so that oil could be sent to Japan did not stir enthusiasm for the venture in Californian hearts. While the Department of the Interior argued that by the time Lease Sale 53 was producing, the glut would have disappeared; Californians remained unconvinced. The crescendo of dissent grew through 1980 and finally in the fall Secretary Andrus threw in the towel and withdrew the four northern basins and the 30 most controversial tracts on the northern edge of the Santa Maria Basin.

The situation was complicated by a parallel OCS development that would have serious implications for California. During 1979 the Department of the Interior prepared and circulated a comprehensive five-year OCS schedule for the entire coast of the nation, in which California would be a major participant. This schedule was the first that had been developed pursuant to the OCS Land Act Amendments of 1978 and it proposed a radically faster pace of development, particularly in frontier areas. While the Alaskan coast was to bear the brunt of this new program, with the entire Alaskan coast offered for lease over a five-year period, California's OCS activity would accelerate as well.

So, in 1980, California, with Alaska and a number of environmental groups, sued the Department of the Interior to withdraw its five-year plan on the grounds that it failed to meet a number of requirements of NEPA, the OCS Land Act Amendments, and other statutes. The basic contention was that it would be impossible to gather sufficient environmental information or to assess the environmental impacts of each sale adequately in the short time proposed between sales, so as to have useful information in time for making intelligent decisions concerning future sales in each area. The new five-year schedule further galvanized activists and the state against what was believed to be a federal program of enormous threat to the state, its economies and resources. California et al. won that suit in the District Court of Appeals in October 1981, once more throwing the national OCS development plan into question.

If California found the Carter administration difficult to convince on OCS development, it was nothing compared to the arrival of Ronald Reagan. Watt's very first public policy statement was on offshore drilling off California, when on February 27, 1981 he announced that he was reversing the Carter administration decision on OCS Lease Sale 53 and re-proposing the four northern basins for the sale that was to take place May 31. But he noted that his decision would not be final until after a 60-day comment period. For the first fight, Watt couldn't have picked a more battle-hardened or wily adversary. Cloaking himself in the holy mantle of state's rights, Governor Jerry Brown took up the challenge. It was well known that the U.S. Geological Survey (USGS) estimated that at least 75 percent of the oil and gas to be found in the lease area was in the Santa Maria Basin that already was going forward. California portrayed itself as the victim of an arbitrary federal giant bent on getting the last drop of oil at any price. Watt's move established his reputation as a politically naive idealogue who shot from the hip. Articles appeared in most of the major papers across the country depicting Watt's action as a capricious betrayal of a federal commitment. This time a letter from 30 members of the California delegation went to the president, and the chairman of the state Republican Party added his voice to the clamor. Hearings popped up on Capitol Hill with plenty of television cameras where California Congressmen expressed their outrage at this trampling of states' rights. The Reagan reputation for honoring the individual and the states against big, bad federal government took its first bruise.

When California sued this time, it sued under the consistency clause of the CZMA stating that the Watt proposal was in violation of its Coastal Zone Management plan. And this time, it was joined by nine states and 11 cities and counties, including North and South Carolina, Florida, and even Alabama. A hot nerve had been touched in the states' rights muscle, and a powerful message was being sent to the administration elected to protect Americans from this kind of arbitrary federal action.

Environmentalists found themselves in an unusual and for some an uncomfortable position. Traditionally, the environmental community pressed for increased federal controls over natural resources on the grounds that state and local governments were too vulnerable to political pressures to be entrusted with authority over resources as valuable and threatened as those of our coast. But in this case, they stood shoulder to shoulder with the state and town governments that reflected the values of a constituency with an unusually high environmental consciousness.

In August, Judge Mariana Pfaelzer ruled that Lease Sale 53 changes did indeed violate Section 301(c)(1) of the CZMA, and Secretary Watt announced that he had decided against including the four northern basins and the 31 disputed tracts of the already leased Santa Maria Basin. It was a difficult loss of face for the Secretary and a very important political lesson. Since then Watt has not moved forward on Lease Sale 73 and it is suspected he is waiting for a more malleable governor after 1982.

During 1981 another high drama developed in the ongoing saga of the definition of directly affecting in the consistency clause. You remember that this had been left at a standoff at the pass between California and the Department of the Interior as Carter and his entourage packed their bags. In February, Watt sent to Commerce Secretary Baldridge an interesting letter in which he stated that a new policy on the consistency problem was an urgent priority, and he urged Baldridge to move forward quickly to produce regulations for the directly affecting clause that would rule out its applicability to preleasing activities. In other words, Watt intended to defang California before a court decision by writing a regulation explicitly preventing states from objecting to the leasing of specific tracts. Baldridge dutifully complied and just before the judge took the Lease Sale 53 lawsuit under consideration, the regulation stating that directly affecting applies only to those activities that have a physical, concrete impact on the coast was enacted, explicitly stating that the consistency clause did not apply to preleasing activities.

The Judge, however, ignored this new regulation, and sided with California. But the consistency fight had just begun with the appearance of the new Watt inspired regulation. In 1980 the Coastal Zone Management Act had been considerably amended, and one amendment that environmentalists had bitterly opposed gave Congress the authority to overturn any regulation promulgated under the CZMA. This kind of authority amounts to an extra heavy club held over any agency's head in addition to the already potent threat of the power of the purse that Congress traditionally wields. The amendment was the inspiration of Ed Forsythe (R-N.J.), who was the Committee's proponent of faster OCS energy development.

Shortly after Commerce's new regulation appeared, Congressmen Gerry Studds (D-Mass) and Joel Pritchard (R-Wash) and Senators Hollings (D-S.C.) and Weicker (R-Conn) introduced a resolution to force the Department of Commerce to withdraw the regulation. Thus, the states' rights issue was carried to the Congress, where it became a *cause célèbre.* Senator Fritz Hollings is a traditional southern populist, not a flaming liberal. Joel Pritchard is a cautious Republican who occasionally takes on controversial issues but views himself as the architect of compromise. The Congress had only 60 days to pass the resolution from the time of the regulation's promulgation--something akin to achieving the speed of light by congressional standards. But hearings were quickly held in the House Merchant Marine Committee, the resolution passed out of the Oceanography Subcommittee within a week and sites were set on the full committee vote. A number of committee staff commented afterwards that they had never seen such intense lobbying on both sides of any issue before the committee. Robinson West, Assistant Secretary for Policy, Budget, and Administration of Interior met with some members as many as three times. The Coastal States Organization, an association of coastal-zone state managers, alerted their governors' offices, and the calls and letters from state governors poured in, in favor of the resolution. Enviromentalists really took a back seat, though it was a critical environmental issue, to the passionate appeals being made on behalf of states' rights. Fishing groups got involved as did recreation associations. When the smoke cleared, the resolution passed by a comfortable margin of five votes.

Action in the Senate was more difficult, however, since Committee Chairman Bob Packwood (R-Ore) had already challenged his party leader on AWACS and some other issues and was reluctant to take on the president again. But as time was running out, a hearing/mark-up looked as though it would be scheduled just in time to make it to the floor, and suddenly the Department of Commerce withdrew the regulation the first week of October, conceding at least a temporary defeat.

The consistency fight was one of those obscure bits of legislative drama that marked an enormous power struggle over very fundamental issues. How much should states sacrifice for the sake of the national energy crisis? This question leads into a labyrinth of murky factual information. Will very costly OCS development in frontier areas achieve the quickest, most efficient path to energy self-sufficiency? Environmentalists can trot out reams of data to show that simple energy conservation has been the quickest, cheapest road to energy self-sufficiency and will continue to be so if free market conditions are allowed to persist. Energy industry representatives likewise can show the importance of every gallon of OCS oil to energy self-sufficiency.

This struggle to control national policy in OCS development is far from over. The Reagan administration is merely biding its time before making another foray on the consistency issue. It certainly has not conceded defeat. How far are the states willing to carry the cause? What will be the impacts on the environment if in the end the states lose the authority to force the feds to delete tracts or, indeed, whole basins? How much energy will we really lose if state concerns are met? These are all questions without clear answers as yet.

CONCLUSIONS: COMPARISONS WITH
GEORGES BANK AND ALASKA

The California experience illustrates the contrasts between the experience in a state with a progressive view towards conserving its natural resources, including an enviromentally conscious citizenry, and states like Massachusetts that watched while a resource of comparable economic value, the Georges Bank fishery, was leased away under its nose with very little protest. An extremely conservative governor in Massachusetts and some not very environmentally oriented statehouses in the rest of New England were never stimulated to action, because the fishing community most affected had a very difficult time organizing. It was too small, too diverse, without resources to take days off from the boat and go up and lobby in the state capitals. Finally, a crucial tactical error was made by the environmental community to put almost all its resources into a lawsuit by the Conservation Law Foundation. While this did achieve delay of the sale, ultimately the suit lost, and there was not sufficient emphasis put into basic grassroots organizing to stop the sale. Since all the affected states with CZM programs certified that the Georges Bank sale was consistent with their coastal plans, there was little recourse for local activists.

The California experience raised the consciousness of a number of states concerning their rights and OCS development. That nine states joined California on its consistency lawsuit marked a revolution in attitude toward offshore drilling. North Carolina became alarmed at the proximity of drilling and sued to force the Department of the Interior to move farther off its magnificent beaches. Alaska, which will bear the brunt of the new five-year schedule, learned much from California. Unwilling to take on the president so directly, Alaska has been involved in tense negotiations behind the scenes with the Department of the Interior over important threatened areas.

An unfortunate irony is the incredible media blitz that California was able to focus on itself. It tended to obscure a far more critical struggle being waged over the Alaskan OCS, where the stakes are far higher. While California captured all the press, Secretary Watt incorporated into yet another, faster five-year OCS schedule in 1981 a new system of leasing that would do away with tract by tract environmental assessments as well as speed up the lease offerings. Entire basins would be offered at up to 1 million acres at a time with only a general environmental impact study for the entire sale. The impact of such a system on California would be very serious, but on Alaska it would be catastrophic.

So little is known about the living resources of the 30,000 miles of pristine Alaskan coast and its waters that it would be impossible to assess the impacts of drilling in much of the OCS. Furthermore, arctic weather and ice conditions present a true frontier to the energy industry, unlike even the North Sea experience. In arctic climates recovery from oil spills would take at least twice as long as in more temperate zones. With a far more conservative government and constituency, resistance to this unfettered exploration has been weak despite an active lobbying campaign by native subsistence users of coastal resources, fishermen, and environmentalists.

Opening up Bristol Bay, the greatest salmon estuary in the world, or the Beaufort Sea, crucial habitat for the bowhead, most endangered of all whales, as well as many other areas of rugged beauty and incredible abundance of birds and marine mammals, will likely alter forever the nature of Alaska's coastal ecosystems. Finding a context that works in which to evaluate these momentous tradeoffs is extremely difficult. Conflicting national interests and conflicting state versus national interests have a way of getting resolved, not through careful evaluation of the coastal benefits, but according to which interest group can wield the greatest political muscle. For California, Alaska, Georges Bank, and other areas of tremendous coastal riches, the struggle has really just begun; in a few short decades, when all the oil is gone, what will be left of the renewable living resources that inhabit the outer continental shelf?

REFERENCES

Much of the information for this chapter comes from my personal experience in working on marine sanctuaries, OCS legislation and administrative issues during the last three years. Most of the facts have been presented at Congressional hearings where I have been a participant and observer.

Baldwin, Malcolm, and Page, James, eds. 1970. *Law and the Environment.* New York: The Conservation Foundation, Walker and Company.
California Coastal Commission, *Policy and Legal Positions of*

the CCM Regarding Lease Sale 48 Consistency Mediation, September 7, 1979.

California Comments on Lease Sale 53 to Secretary of the Interior, incorporating agency and county comments, Governor's Office, Sacramento, California, December 29, 1980.

California Comments and Findings on Lease Sale 53, incorporating agency and county comments, April 7, 1981.

California v. James G. Watt, Case No. CV 81-2080, Opinion, August, 1981.

Healy et al., eds. 1978. *Protecting the Golden Shore.* Washington, D.C.: Conservation Foundation.

Santa Barbara County. 1978. *Nomination of the Santa Barbara Channel as a National Marine Sanctuary.*

Sollen, Bob. 1978. "The Santa Barbara Story," *Not Man Apart.* San Francisco: Friends of the Earth.

PART II

THE
STATES'
PERSPECTIVE

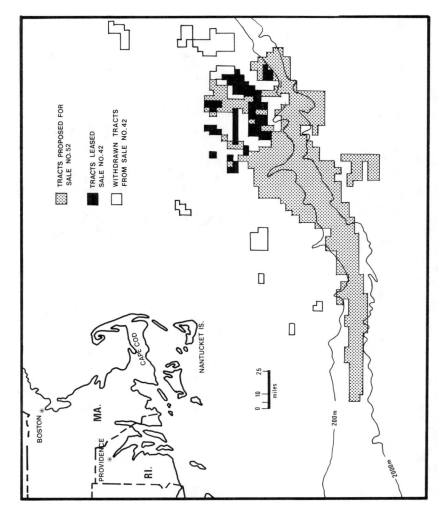

Lease Sale Number 52 and Number 42 of Georges Bank

2
The Search for
an Ocean Management Policy:
The Georges Bank Case

Charles S. Colgan

In November, 1974, a little more than a year after the Yom Kippur War and the subsequent Arab oil embargo had demonstrated the serious energy supply problems that the United States was facing, Secretary of the Interior Rogers C.B. Morton held a press conference in Washington to announce a vast increase in the department's leasing of outer continental shelf (OCS) acreage. Morton said that up to 10 million acres per year would be offered in the new leasing program. Included in the increased acreage would be leasing in the Atlantic and other frontier (never-before drilled) areas. The program included the opening of the Georges Bank, a submerged plateau lying off the coast of New England, to oil and gas exploration. The first lease sale for the Georges Bank, designated Lease Sale 42, was scheduled for August, 1976.

On July 24, 1981, at 1:34 A.M., the semisubmersible drilling rig *Zapata Saratoga,* under contract to Shell Oil Company, lowered a 12 inch drill bit to a temporary guide base on the ocean floor, some 85 meters below the surface of Block 410, 140 miles southeast of Nantucket Island. Twelve hours later and 30 miles northwest, the drilling rig *Alaskan Star,* working for Exxon, also spudded its first well. Almost seven years after Rogers Morton's announcement, and five years after the first Georges Bank lease sale was originally scheduled, oil and gas exploration began on Georges Bank.

The question of why it took seven years from the first official announcement of OCS leasing on Georges Bank and the actual beginning of drilling is one that has been frequently raised. For some, particularly the oil industry, the question is asked in terms of "Why did it take so long, given the critical energy needs of the United States?" For others, including the New England fishing community that relies on Georges Bank for its livelihood and produces some 17 percent of the world's fish from the nutrient-rich waters of the bank, the question is asked in terms of "Why so soon?"

Whether the time lapse between intention and event was too long or too short is an issue that cannot be settled until the oil industry finally completes its efforts on Georges Bank, either by drilling a number of dry holes that force abandonment of future exploration efforts, or after a commercial-sized find of oil and gas has been located, developed, and produced. The end of oil operations could come in five years or 30 years.

But the period between 1974 and 1981 was part of the most critical era in the development of national ocean policy in the United States. Beginning with the Clean Water Act in 1970, and proceeding through the Coastal Zone Management Act, the Marine Protection, Research, and Sanctuaries Act, the Fisheries Conservation and Management Act, and the OCS Lands Act Amendments, the federal government undertook to define goals, policies, and means for management of the nation's coastal and ocean resources.

The passage of these laws by Congress signaled an awareness that the coasts and oceans were resources of considerable significance. The laws were also signs that the conflicting needs of fishery resource conservation and development, the development of new energy sources, and protection of the marine environment were going to have to be made on a case-by-case basis. And nowhere were these conflicts more apparent, nor the policy problems involved in such a case-by-case approach more difficult to resolve than in the case of Georges Bank.

THE SEARCH FOR OIL ON GEORGES BANK

The Georges Bank is an oval shaped area of the Atlantic OCS that lies offshore Massachusetts, just south of the Gulf of Maine. It is about 150 miles long and 60-80 miles wide. Georges Bank is somewhat shallower, and contains warmer water than the ocean surrounding it. Because of its shallowness it is characterized by complex current patterns that combine a semiclosed clockwise circulation pattern (gyre) around the edge of the bank, with strong diurnal tidal currents. This combination allows significant mixing of water and provides the opportunity for nutrients that lie near the bottom to be brought to the surface and thus made available to the plankton life on which the large commercial fishing industry of New England depends for food. The bottom of Georges Bank is characterized by environments that range from sandy, suitable for scallops and ocean quohogs, to rocky canyon areas where large lobster populations are found. The combination of all these characteristics make Georges Bank the richest fishing ground for its size in the world.

It is little wonder that the prospect of oil exploration, only a few years after the Santa Barbara blowout, which had been a major catalyst for the environmental movement of the 1970s, was not viewed with universal approval in New England.

In 1975 two events occurred that were to define the issues that would be debated through the next four years, leading up to the actual conduct of Lease Sale 42 on December 18, 1979. First, the issue of who owns the OCS was resolved by the Supreme Court (*United States* v. *Maine*). In 1969 the state of Maine asserted that colonial statutes conferred ownership of the OCS to the individual colonies and their sucessors the states, rather than to the federal government. The state actually sold oil exploration rights to an area of the Gulf of Maine to the King Resources Company. The federal government filed suit against Maine, which was joined by other Atlantic coastal states, and no exploration actually took place. The suit was not settled until 1975 when the Supreme Court ruled that ownership of the continental shelf beyond three miles was vested solely in the federal government.

The second event was consideration by Congress of a series of changes in the basic law governing oil and gas activity on the OCS, the 1953 OCS Lands Act. That law established the basic system under which oil companies purchased exploration rights at sealed-bid auctions held by the Department of the Interior, and assigned overall regulatory responsibility to the department. In the fall of 1975 the House Select Committee on the Outer Continental Shelf held a series of hearings on the Pacific, Gulf, and Atlantic coasts. At the hearings held in New London, Connecticut and Boston in September, the concerns raised by the planned leasing of Georges Bank were first voiced.

There were six basic themes voiced by participants at the hearing. A concise summary of the concerns of New Englanders was provided by Massachusetts Governor Michael Dukakis:

Every aspect of the leasing system is outdated. It provides almost no opportunity for state involvement in decisions that are integral to our future development; it provides for little or no access to the information and data we need to plan and manage potential impacts; it provides no compensation, planning assistance, or up-front monies to the coastal states; and it does not allow us to determine the scope of a resource before allowing the oil industry to develop that resource. [Dukakis, 1975a]

The consensus among a majority of participants in the hearing was that before the federal government could begin the expansion of OCS leasing it was contemplating, a number of changes were needed in federal law. These changes are covered below.

The Need for Expanded Public, Particularly State Involvement

The state governments were in the process during 1975 of developing their coastal-zone-management programs as authorized by the 1972 Coastal Zone Management Act and as a result has developed both an awareness of and a staff capability to address the demands that might be put on onshore and offshore resources if OCS development occurred. As a result the states were insistent that their voices should be heard in federal decision making. The point was made by Governor Dukakis: "If there is one issue that most of the Atlantic states agree on, it is the need for us to participate in the design and implementation of OCS programs. We want a 'guaranteed partnership' written into law which would mandate participation and full consultation" (Dukakis, 1975b). The point was reinforced, and the notion of a guaranteed partnership given substance by Governor Longley of Maine:

Under current law, the States have no power to affect basic decisions such as which tracts are leased or the approval of development and production plans. This is an untenable situation. The answer is simple: Write a guaranteed partnership into law. Regional advisory boards and individual governors should be guaranteed by law that their formal recommendations on any aspect of OCS leasing and development will be accepted unless the Secretary [of the Interior] finds them inconsistent with the national interest. [Longley, 1975b]

The Governors were not the only critics of the lack of opportunity for public input to the OCS decison-making process. A fisherman who appeared at the hearing also noted the problem: "The major fault which I see with the present leasing procedure is an unsatisfactory opportunity for public input into the decision making process. In particular, of course, I am concerned with the opportunity to obtain consideration for the needs of the commercial fisheries" (Allen, 1975).

The Need for More Careful Planning before OCS Development Takes Place

There was substantial distrust expressed that the federal government had clearly thought through the implications of their OCS program, and that the oil industry was insufficiently motivated to consider the needs of coastal states and fisheries:

> If we have learned anything from the last 30 years it is that profit alone cannot guide development. . . . There must be careful planning and management of this problem and development of these resources. Planning may not be our salvation, but it can make the difference between uncontrolled confusion and an orderly, disciplined response to the energy crisis. . . . [Dukakis, 1975b]

Suspicion of the oil companies was something of an endemic response in New England, which had historically been heavily dependent on oil for a major part of its energy supplies, but being at the end of the pipeline had paid the highest prices for oil in the United States. Before the Arab oil embargo dramatically reversed the situation, imported oil was substantially cheaper than domestically produced oil, and the oil import quotas had served to keep out the lower-priced oil. New England had fought long and hard to remove the quotas, but oil industry support for the quotas had kept them on and prices in New England high. Even Maine Governor Longley, elected as a candidate pledged to bring business practices to Government, reflected the skepticism of the oil industry:

> [I am] exceedingly uncomfortable with the influence and the impact of the major oil companies in this country and their complete failure to solve the needs

and the national security problems of this country. So as Governor of Maine, I am all the more uncomfortable that the major oil companies are being given too much freedom in the present development approach. [Longley, 1975a]

The expressed need for better planning was actually concerned with two aspects of the forthcoming development. First was the need to better understand the offshore ecosystem, and second the need to plan for onshore impacts.

Environmental Studies to Determine the "Baseline Conditions" Prior to Drilling

Beginning in 1974 the Department of the Interior had undertaken a series of studies, through the Bureau of Land Management (BLM) (the agency within the Department of the Interior responsible for conducting the lease sales) to determine what the condition of the ocean systems were prior to drilling. The hope was that by gathering this baseline data that impacts from drilling could be identified if they occurred. These studies for Georges Bank were still being planned at the time of the hearing, but the need for them as a prerequisite to drilling was expressed by Steven Weems, the Maine OCS coordinator:

. . . I want to reemphasize the point about how to protect the fisheries. The so-called environmental baseline studies . . . really must be initiated and a good portion of them--say perhaps a minimum of one year--completed before making any fundamental decisions in frontier areas. . . . We need to understand more about the potential impacts of leasing an area before taking the plunge. [Weems, 1975]

Onshore Impacts in New England

Offshore oil development requires support from a variety of onshore facilities, ranging from the service bases that supply materials to the rigs, to the platform construction yards where the giant steel production platforms are fabricated. Accompanying the location of such facilities came rapid population growth (often in small coastal communities), rising demand for government services, and wrenching social change. In areas such as Scotland, the Shetland Islands, and some areas of Louisiana this has become a major concern (Baldwin and Baldwin, 1975).

Speculation about the possibility of future onshore facilities was occurring throughout New England. One individual was actively attempting to develop a service base in Searsport, Maine and a mysterious New York promoter was looking to establish a base on Nantucket:

> Efforts are now being made to transform Nantucket into a helicopter base for offshore oil rigs, and a logistical support base for the rigs. . . . [It] is not difficult to predict the results of [such a proposal]. The Nantucket Airport and surrounding property would be covered by a new and major motel, hangars, repair and maintenance facilities, parts and material warehouses, and other storage facilities and catering services for the rigs. Presumably, R and R facilities would be needed, although it is hard to imagine creating sufficient of these on Nantucket to satisfy the thousands of workmen who may ultimately be involved. The impact of trucks and equipment on the narrow, winding roads of Nantucket can easily be foreseen. The population of the Island, now lacking in the labor skills which the oil rigs require, inevitably will expand several times over. Taxes will inevitably increase for more schools, police, fire protection, etc.,--taxes which will largely be borne by existing property owners if past history of real estate development is any criterion. Over one-hundred years ago the discovery of oil at Titusville wrecked the whale-oil based Nantucket economy. Is offshore oil exploration and production going to wreck the present economy, and replace it with a transitory one that will inevitably destroy the island and the quality of life as we know it today? [Gifford, 1975]

The fears of such development in New England coastal communities engendered demands for a number of federal actions:

> [Information, including] the extent of oil and gas resources, and the facilties both onshore and offshore needed to extract, transfer and process these petroleum resources . . . [are needed by] local and state governments . . . to manage offshore and coastal area resources effectively.

Based on the likely speed and intensity of development activities, it is clear that potentially affected states need financial assistance to [citizens, communities, and states which must bear the direct and indirect burdens of offshore oil and gas development. Significant federal funds should be made available to prepare economic development plans and impact assessments. A grant or loan program is needed to defray the high initial costs of developing public infrastructure and services, as well as Federal grants to reimburse state and local governments for any net adverse impacts. [Dukakis, 1975b]

Several Changes in Leasing and Management Procedures

In addition to increased state involvement and planning, several basic features of the OCS management system were singled out for change in order to provide better assurances that the environment, particularly the Georges Bank fishery would be protected. There were four major changes that were recommended:

1. After public hearings and in consultation with coastal states the Secretary of the Interior should have the authority to modify the development and production plan in order to avoid serious, environmental, or economic, or safety implications. . . . [I]f at any time there is evidence that certain other resources, such as commercial fisheries, are being seriously impacted, there must be a process for amending the development and production plan. [Dukakis, 1975a]

2. An Environmental Impact Statement and an economic impact analysis on both offshore and onshore impacts must be completed *before* approval of an OCS Development and Production Plan. [Dukakis, 1975b]

3. Holders of leases or rights-of-way on the OCS should be strictly liable, without regard to fault and without regard to ownership of any adversely affected lands, structures, fish, wildlife, or biotic or other natural resources for all damages, including clean-up costs, which result from their activites. To assure responsible development, there should be no limitation on the amount of liability. A federal revolving fund, financed by OCS revenues should be established to insure prompt payment of damages. [Longley, 1975b]

4. The Secretary of the Interior . . . should review existing safety and environmental regulations and promulgate new improved regulations. Such regulations should require the use of the best available technology whenever failure of equipment would have a significant effect on public health, safety or the environment. [Dukakis, 1975b]

The comments of the states reflected a desire for changes in the basic federal-government processes of OCS management, and were driven primarily by a desire for greater state involvement in the decision-making process. The New England fishing industry expressed concerns over both the fundamental wisdom of leasing for oil on Georges Bank, but also over specific problems that the fishermen expected to encounter if oil operations did begin.

. . . [P]roblems being considered [by the National Federation of Fishermen] include locating and protecting pipelines; debris; hangs; competition for onshore facilities; pollution; and distinct changes in social and economic patterns in areas and communities where commercial and sports fishing are now the major offshore activities. A further, longer term issue is the comparative values of, say, 25 years of oil and gas exploration and development from an area versus protein and pleasure from fisheries in the same area over a longer period: the question of how best to manage for maximum benefits of renewable and nonrenewable resources of the same area. [Sloan, 1975]

The 1975 hearings were not all skepticism, however. There were several speakers who exhibited no qualms whatsoever about the imminent oil leasing on Georges Bank, foremost among them being the governor of New Hampshire, Meldrim Thomson:

We face this winter in many of our Eastern Seaboard States the worst shortage of natural gas in our history. Ahead lies the grim prospect of unemployment reaching its highest rates since the Great Depression, due primarily to our shortages of energy. . . . Fortunately, despite myriad attempts by Congressmen and Governors to delay or prevent the extraction of fossil fuels from our continental shelves, the Department of the Interior, with the firm backing of the President, has pursued a vigorous policy of early production of oil and gas on the shelves. [Thomson, 1975]

Governor Thomson was insistent that the need for energy was so great that the states should have no say at all in OCS decision making:

> [I]n the search for and the development of energy I am a nationalist. I believe firmly that energy is as much a national matter as the general welfare and common defense. . . . Beyond the three mile limit I do not believe the Governors should have anything to say about the exploration and development of oil. [Thomson, 1975]

Governor Thomson was joined in his plea for faster action on OCS leasing by several groups of New England businessmen, including the (Massachusetts) South Shore Chamber of Commerce:

> We must once again stress emphatically the need for the immediate commencement of offshore drilling on the continental shelf area of Georges Bank. . . . New England desperately needs energy--for without adequate supplies of reasonably priced energy our economy can only continue its downhill slide. The Council on Environmental Quality . . . in a report in April, 1974 estimated 75,000 additional jobs would be created in our region by 1985 with outer continental shelf development. WE NEED THOSE JOBS. [Frazier, 1975]

The 1975 OCS hearings served to provide a forum for an airing of the major positions regarding OCS development on Georges Bank, but the hearings were not sufficient to persuade Congress to pass H.R. 6218, that would have provided most of the changes requested by Maine, Massachusetts, and others.

The imminence of leasing was lessened somewhat by the announcement that Lease Sale 42 would be postponed to the spring of 1977. Following the call for nominations in June, 1975 at which 1,927 tracts were suggested for leasing by the oil industry and 206 tracts were eventually selected by BLM for study, an environmental impact statement was prepared, with public review and hearings scheduled for late 1976.

While the Draft EIS was being prepared in 1976, Congress again took up revisions to the OCS Lands Act, but continued opposition from the Ford administration prevented passage of any changes to the OCS system except for amendments to the Coastal Zone Management (CZM) Act. These changes, embodied in amendments passed in July, 1976 did address one of the major concerns of the governors: funding to plan for and mitigate the impacts of the OCS energy development. The Coastal Energy Impact Program (CEIP) was established to provide a combination of grants and loans to coastal states that allow planning. In some cases, the CEIP provided funds for the construction of facilities to provide the public services required by OCS development. Congress provided a large appropriation to both the grant and loan pools of the program, and over the next few years the states, particularly in the Gulf of Mexico, California, and Alaska would utilize CEIP funds to address a wide variety of energy facility impacts.

In addition to establishing CEIP, Congress also clarified the federal consistency provisions of the CZM Act to make it explicit that states with approved coastal zone management programs were entitled to review the federal permits required of oil companies prior to commencing drilling. This clarification, though little noticed at the time, would prove to be critical to the way the permitting of exploratory drilling was conducted after the lease sale had been held.

The major event of 1976, however, was clearly the publication of the Draft Environmental Impact Statement (DEIS). Prior to the DEIS, oil leasing on Georges Bank was widely perceived as an event set sometime in the future, that was sufficient to cause concern but not alarm. The EIS, published in October, provided the first widely available analysis of oil activity on Georges Bank. At the public hearings that were held in December in Boston and Providence, the discussion began to focus on the specific details of offshore oil that would have to be confronted.

The DEIS consisted of four volumes that were, according to the requirements of the National Environmental Policy Act, to provide as complete a picture as possible of the Georges Bank environment, the details of the proposed lease sale, and the anticipated impacts both onshore and offshore. The measures that the Department of the Interior proposed to utilize to mitigate any adverse impacts were also to be set out.

Reaction to the DEIS came from many of the same parties that had participated in the 1975 hearings, notably the states. The DEIS also drew the attention of several major environmental groups. Reviewers of the DEIS could be divided into those who supported oil development on Georges Bank without any reservation and those who supported oil exploration in concept, but had significant reservations of one form or another. In addition, the DEIS also brought forward statements of opposition to leasing either entirely or in the immediate future.

The proponents' arguments for leasing were for the most part similar to those made earlier. Jobs and energy were seen as the major benefits to New England, benefits that should neither be foregone nor delayed. Added to these arguments, made primarily by New Englanders, were the arguments of the oil industry that sent representatives for the first time to defend the record of their industry and assure New England that there was no need to worry:

> Unlike the Louisiana area, the Atlantic outer continental shelf does not have the geologic history required to produce high subsurface pressures. Geologically, the Atlantic OCS is the safest region in domestic offshore waters for exploration and production of petroleum. [McAuliffe, 1977]

> In offshore operations, improper disposal of wastes such as drill cuttings poses the only potential threat to the environment. We have taken positive steps to avoid discharge of any drilling fluids or cuttings. Safety devices such as blowout preventers and down hole safety valves are already required by federal regulations, and these we have strictly adhered to, often at great expense to the operator. We have learned to improve the safety of our OCS operations by burying our undersea pipelines, establishing proper waste disposal procedures, and instituting emergency cleanup and containment procedures. . . . For the record, worldwide drilling and production activities have contributed no more than 2 to 4 percent of all the oil released into the oceans. [Matthews, 1977]

The arguments that OCS development on Georges Bank were inherently safe and that New England's needs for employment and energy were sufficient to justify the lease sale were not convincing to other participants at the DEIS hearing. Two Massachusetts legislators expressed their doubts:

> I . . . recommend delaying the sale until the technology of new equipment to provide increased environmental protection has been developed. I am not persuaded that such technology exists at the present

time. . . . For example, there exists no booming equipment which can handle the 4 or 5 knot current in Buzzards Bay Canal. [Pines, 1977]

There is something very wrong with our priorities when we are consumers of a product that is blatantly as wasteful as an automobile that can go 125 mph with a ten mile per gallon consumption rate, or support a supersonic airplane which consumes even more gasoline, while making more noise, polluting more of the atmosphere for fewer passengers for the purpose of getting somewhere faster. [Kendall, 1976]

The analysis in the statement convinced some reviewers that the lease sale entailed too many risks, and others that not enough was known about the impacts to be able to make a decision to go ahead:

From your statement one learns that oil spillage will be unavoidable, possibly in catastrophic amounts and that the short term effects of oil pollution on marine life, vegetation, commercial fishery resources, recreation, water, and air quality will be adverse and in some cases irreversible. And, perhaps the greatest concern of all, your reports states in paragraph after paragraph, that the long term effects are basically unknown. . . . I consider these known and unknown environmental and economic hazards too great a risk . . . to support OCS Sale #42. [Kendall, 1976]

[D]rilling on Georges Bank at this time amounts to playing Russian roulette with our fishery resources because we do not know the location of some of the spawning areas with any precision. We *do* know that Georges Bank produces one eighth of [the] world's offshore fish catch; we *do* know that fish larvae are found near the surface where they would be hit by any spilled oil; and we do know that they are much more sensitive to oil than adult fish and can be killed by concentrations on the order of one part per million. The EIS concedes our ignorance. . . . Until the effects of oil drilling on the Massachusetts coastline and the fishing industry are known, we believe that no leasing should proceed. [Sierra Club, 1976]

Between the proponents and opponents was the larger group that emphasized the need to implement specific measures prior to allowing any drilling. The members of this group had their doubts about the information in the DEIS and the impacts of oil operations also: "The EIS makes clear how little we know about the possible effects of allowing large scale oil and gas production efforts in the northwest Atlantic. What we do seem to know is not comforting" (Studds, 1976). But this group focused attention primarily on recommendations for mitigating measures to be added to those that BLM indicated would already be implemented.

In the DEIS, the department argued that existing regulatory measures, coupled with regulations developed for Georges Bank would be sufficient to assure basic protection of the environment. The measures that were to be implemented included:

1. USGS Operating Orders: These regulations formed the heart of the department's oversight and control of actual drilling operations. The U.S. Geological Survey set forth detailed requirements in such areas as casing, cementing, blow-out preventers, well abandonment, and construction of mobile and fixed platforms;

2. The Environmental Studies Program: The BLM Environmental Studies Program, which was established to gather the baseline data prior to drilling was held out as providing the information necessary to determine whether any adverse impacts had occurred;

3. Pipeline regulations: In addition to managing the leasing process for the drilling operations, the BLM also has responsibility for leasing pipeline rights-of-way. In the DEIS, the Department of the Interior noted that subsequent analysis and regulation of pipelines would assure minimum impacts from pipelines;

4. Inspections: The Geological Survey has the power to conduct routine and spot inspections to monitor compliance with operating orders and other regulations;

5. Development Plans: Under a regulation first published in 1975, operators are required to submit detailed development plans after a find is made but prior to the installation of any development, production, or transportation facilities;

6. A prohibition of the disposal of oil-based muds and oil-contaminated drill cuttings;

7. Oil Spill Contingency Plans: A contingency plan for the oil operations would have to be prepared as a supplement to the National Oil and Hazardous Materials

Contingency Plan. The DEIS generally discussed the existing technology of oil-spill cleanup, but could not specify the equipment and procedures that would ultimately be implemented. [Bureau of Land Management, 1976]

The regulations that were developed specifically for Georges Bank were to be contained in lease stipulations. As stipulations attached to the actual leases, the department argued that it had maximum flexibility in designing site- or area-specific regulations. The major environmental stipulations proposed for Sale 42 included:

1. A requirement that cultural resource surveys be conducted to determine if there were any archaeological or cultural sites of concern that could be affected by the drilling;
2. A requirement that if pipelines are determined to be environmentally preferable and economically feasible, that they would be required for transportation of oil and gas ashore;
3. That operators would have to file a "Notice of Support Activity" prior to beginning exploration. This notice would be provided to the coastal states and would detail the location, size, and general activities expected at any onshore support bases established for exploration;
4. A requirement that drill muds be either shunted to a depth of 20-50 feet below the surface, or if necessary the muds were to be barged offsite for disposal;
5. That if oil or gas were to be produced, the department would require unitization of field operations, that is, it would require development of an oil and gas field as one unit, without respect to the boundaries of tracts;
6. Equipment that could be lost overboard of the rigs or the supply boats and that could snag and damage fishing gear would have to be marked with the owners name in order to facilitate claims against the operator for damages.

The final stipulation proposed (actually listed as Stipulation 2) was one dealing with the possible modification of operations in identified biologically important areas. Because it was considered central to the protection of the fishing industry, it is quoted here in full:

7. When an area or resources has been identified as biologically important by a committee composed of designated representatives of the Bureau of Land Management, the U.S. Fish and Wildlife Service, U.S. Geological Survey, the National Marine Fisheries Service, the Environmental Protection Agency and representatives of the affected states, the Supervisor may give written notice that the lessor is invoking the provisions of this stipulation. The first definition of such areas will take place before exploration starts in the lease sale area. . . . The lessee shall, upon receipt of such notice comply with the following notice:

Prior to any drilling activity or the construction or placement of any structure for exploration or development of lease areas . . . the lessee shall conduct environmental surveys, as approved by the Supervisor after consultation with the Committee, to determine the extent and composition of biological populations within the area covered by the lease.

Based upon results of the survey, the lessee may be required to (1) relocate the site of such operations so as not to adversely affect the area identified; or (2) modify his operation in such a way as not to adversely affect the area identified; or (3) establish to the satisfaction of the Supervisor, who will consult with the Committee, that, on the basis of the environmental survey, such operations will not adversely affect the area. . . .

This stipulation was the department's major effort to protect the environmental and fishery resources of Georges Bank. It reflected the department's belief that even if insufficient information were available at the time of the lease sale, it would be possible to collect more information later and modify operations as the new information became available. The stipulation also attempted to address the states' concerns about inclusion in the decision-making process by making them members of the committee that would identify the biologically important areas.

But the bureau's list of mitigating measures was clearly insufficient from the point of view of many of the commenters. There were echoes of the 1975 Congressional hearings in calls for an oil spill liability fund, a fishermen's compensation fund, and a second EIS to be prepared at the time of any development and production plan. There were several other recommendations as well:

1. A training program for all offshore oil personnel to familiarize them with the characteristics and operations of the New England fishing industry. [Massachusetts, 1976]
2. The establishment of set corridors for the supply boats traveling to and from the rigs. [Massachusetts, 1976]
3. Deletion of various nearshore and heavily fished tracts in the northwest section of the lease sale area. [Massachusetts, 1976; Maine, 1976]
4. A ban on tankers to transport oil ashore. If insufficient oil were found to justify a pipeline, the field would not be developed. [Conservation Law Foundation, 1976]
5. The assertion of an explicit right to cancel leases if environmental damage was discovered. [Conservation Law Foundation, 1976]

At the conclusion of the DEIS hearings all the major and minor points of view on the oil development of Georges Bank had been aired. Ahead lay the preparation of the Final EIS and the sale itself. But with the turn of the year to 1977 came a new administration in Washington, and a new Secretary of the Interior, former Idaho Governor Cecil Andrus. As the final EIS was being prepared in the winter and spring of 1977, Andrus announced that the Carter administration would support substantial changes in the OCS Lands Act along the line of the legislation that had been defeated in 1975 and that the states had been requesting all along. Two bills, H.R. 1614 and S.9 were introduced and proceeded through another round of hearings in late 1977.

Following completion of the final EIS in April, the Department of the Interior staff prepared the paper work for a secretarial decision. In October, Andrus sent the first proposed notice of sale to the governors for their review. This notice outlined the specific tracts proposed for leasing and the stipulations that would be attached. The issuance of this notice would be required by the OCS Lands Act amendments then before Congress, but Andrus issued it to show his desire to include the states in the decision-making process.

At this point in the Georges Bank process many of the original concerns of the states had been addressed in one way or another. State participation had been expanded through the proposed notice and inclusion on the biological committee. Increased information concerning onshore impacts was to be provided through the development plan and the notice of support activity that the department had indicated it would require. Onshore impacts were to be mitigated by funding from the Coastal Energy Impact Program.

But there were also some omissions that were regarded as very serious particularly by Massachusetts and the Conservation Law Foundation. For Massachusetts the two most important omissions were a fund to pay compensation to fishermen in the event of damages to their gear and to fund cleanup and damage costs in the event of an oil spill. These features were regarded as so important by Massachusetts that staff from the Massachusetts Office of Coastal Zone Management and the Lieutenant Governor's Office spent the majority of their time in the fall of 1977 trying to convince the department that it was empowered by existing law to establish compensation funds from gear damage and oil spills by making them lease stipulations.

In addition, Massachusetts insisted that the department make a commitment to require a development phase EIS and that the Governor's recommendations would be accepted unless in conflict with an overriding national interest (the guaranteed partnership that had been advocated in 1975).

The Conservation Law Foundation did not comment on the proposed notice of sale (since that was limited to Governors), but two of its major concerns were still unanswered: the use of tankers and the right to terminate or cancel a lease.

On December 30, 1977 the department announced that Lease Sale 42 was scheduled for January 31, 1978 in Providence, R.I. The final notice announced an offering of 128 tracts, down from 206 that had been studied in the EIS. For various reasons, 78 tracts had been removed. Some had been deleted on the recommendation of the State Department because of the unresolved U.S.-Canadian boundary on Georges Bank and others had been removed on the recommendation of the states because of conflicts with fishery areas of nearshore locations. Only one tract recommended for deletion was included in the sale notice.

The secretary included a commitment to prepare a development phase EIS, and a notice that the best available technology would be required of all operators. The department indicated that the remaining issues of compensation funds and establishment of a right to cancel leases would be addressed when Congress completed action on the OCS Lands Act Amendments.

Despite the fact that there remained only four major changes to the OCS leasing process uncompleted, the Governor of Massachusetts and the Conservation Law Foundation concluded that leasing should still not proceed. The fear that, despite administration support, the OCS Lands Act Amendments might still fail to get out of Congress was a major factor in this decision. In addition there was concern, especially on the part of CLF, that even if the amendments passed there would be no retroactivity to its provisions if leases had already been sold for Georges Bank. CLF had warned of its concerns in the DEIS hearings: "Only regulations and orders in effect at the time of the lease sale are, by the terms of the OCS Lands Act, enforceable by cancellation. Therefore any orders and regulations of true substance . . . must be prepared prior to the sale" (Conservation Law Foundation, 1976).

So, on January 19, 1978 the Commonwealth of Massachusetts and CLF filed suit in U.S. District Court in Boston, seeking a preliminary injunction to halt Lease Sale 42, then just 12 days away. The arguments were heard over the next three days, and on Saturday, January 28, 1978 the court issued the preliminary injunction. Monday the 30th the federal government sought review of the injunction in the First Circuit Court of Appeals, but that court declined to overturn the injunction. The next day the department issued a notice cancelling the sale.

Massachusetts and CLF won the injunction because they convinced the district court that Cecil Andrus' decision to proceed with the lease sale before passage of the legislation that he was supporting was improper.

> In not waiting for this legislation to be enacted, the Secretary has done less than he should . . . to preserve the natural resource on the Georges Bank which he is bound to do. . . . The whole point . . . of the Secretary seeking this legislation from the Congress as being of the most urgent necessity is premised on his conclusion that he lacks the power to accomplish these ends administratively. Therefore, action by Congress is essential, and it is his refusal or declining to await Congressional action here which the Court believes to be a violation of his duty. [*Massachusetts v. Andrus*, 1978]

The duty to which the court referred was the duty to protect the fisheries which CLF and Massachusetts had argued was inherent in the OCS Lands Act and parallel legislation such as the 1976 Fisheries Conservation and Management Act and the 1972 Coastal Zone Management Act. The court agreed that such a duty was implied by these statutes:

> The OCS Lands Act is the most important statute here applicable. 1332-B of Title 43 provides . . .: "This subchapter shall be construed in such manner . . . that the character as high seas of the waters above the Outer Continental Shelf and the right to navigation and fishing therein shall not be affected . . ." [T]he Secretary of the Interior is directing [sic] as a matter of policy not to affect fishing if it can possibly be avoided. [*Massachusetts* v. *Andrus*, 1978]

In addition to the finding of violations of the OCS Lands Act, the district court also found violations of the National Environmental Policy Act in that the final EIS did not adequately consider the alternative of waiting until legislation was passed before proceeding with the lease sale. The court also accepted the argument that the EIS should have considered transferring management authority for Georges Bank to the National Oceanic and Atmospheric Administration (NOAA) in the Department of Commerce under the Marine Sanctuaries Act.

Finally, the court criticized Andrus for "arbitrary and capricious" actions in proceeding with the lease sale because of the need to fulfill his promise to have a predictable leasing schedule.

> The court notes the emphasis by the Secretary on his having given his word to the industry that having established a schedule not just with respect to Lease Sale 42 out here on Georges Bank but with respect to other options in other sections of the country and with respect to other tracts, he would stick to that schedule. . . . [I]t would in this Court's view be arbitrary and capricious for him to make a decision on the basis of his having promised a sale and an unwillingness to postpone it in order not to inconvenience people who are counting upon the bids being received and opened on the 31st of January. [*Massachusetts* v. *Andrus*, 1978]

With the cancellation of the lease sale the department elected to await the outcome of the legislative process on the OCS Lands Act Amendments and the full review of their appeal by the First Circuit Appeals Court. Nine months after Massachusetts and the Conservation Law Foundation filed suit to stop Lease Sale 42 and three years after the original amendments had been introduced, Congress finally passed the OCS Lands Act Amendments.

These amendments, which included amendments to the Coastal Zone Management Act as well, finally embodied in law virtually all the changes that New England had fought for since 1975. The new law included oil-spill and fisheries compensation funds (Titles III and IV), requirements for coordination with state and local governments (Section 19), and Oil and Gas Information Program (Section 26), requirements for a Five-Year Leasing Program (Section 18), the use of Best Available and Safest Technologies (Section 21), and requirements for a Development Phase EIS at least once in each frontier area (Section 25). The amendments also included modifications to the Coastal Energy Impact Program, including a new grants program to fund the states' participation in the OCS decision-making process.

Following passage of the amendments, the only apparent obstacle remaining to the Georges Bank Lease Sale was the injuction that was still in effect as a result of the Massachusetts-CLF lawsuit. The appeal by the federal government had still to be decided, and in February, 1979 the first circuit handed down a decision that vacated the injunction. But the issue was by no means resolved, for the appeals court made several findings that forced the Georges Bank leasing process into a new round of conflict.

The appeals court basically concluded that since the original injunction had been issued as a result of the Secretary's failure to await the outcome of the legislative process, and since the amendments had now been passed, there was no longer any fundamental issue. The court went on to dismiss any claims that inadequacies in the EIS were sufficient to justify continuing the injunction.

But the appeals court reached a different conclusion concerning the duty of the Secretary of the Interior to protect the fisheries. Referring to the same section of the original OCS Lands Act as the district court had to the effect that "fishing . . . shall not be affected," the Circuit Court noted:

> Plaintiffs have argued that this imposes a duty on the Secretary to see the mining and drilling are conducted absolutely without harm to fisheries. However, it is clear the clause was inserted with no such purpose in mind. . . . The reference to "fishing" was inserted in the bill that eventually became the OCS Lands Act to

show that the United States' extension of its jurisdiction into the OCS was limited to the subsoil and
seabed. The Nation was not, by this Act, asserting
control over the waters of the region. [*Massachusetts
v. Andrus*, 1979a]

The Court did find a duty in the OCS Lands Act with regard to the fisheries, but it was not the duty to protect
the fisheries as a primary goal.

[W]e believe that both past and present versions of
the OCS Lands Act place the Secretary under a duty to
see that gas and oil exploration and drilling is conducted without unreasonable risk to the fisheries.
His duty includes the obligation not to go forward
with a lease sale in a particular area if it would
create unreasonable risks in spite of all feasible
safeguards. As we see it, the question is not whether
the Secretary's task is to put the interest of fishing
"above" everything else, on the one hand, or whether,
on the other, he is mindlessly to lease every square
foot of seabottom at whatever risk to other resources.
Both the previous and amended OCS Lands Act seem to us
to reflect Congress' underlying belief that mineral
development can be so orchestrated with the development and preservation of renewable resources like fish
as to no irreparable harm to them. It is left to the
Secretary to harmonize the interests of the various
resources wherever they impinge on one another.
[*Massachusetts v. Andrus*, 1979a]

On one point, however, the appeals court did agree with
the district court: that the Department of the Interior
should have paid more attention to the option of establishing a marine sanctuary on Georges Bank.

[U]nder the Marine Sanctuaries Act the land use options of the Secretary of Commerce are much the same
as those of the Secretary of the Interior under the
OCS Lands Act, the management objectives are different. It is thus possible that different environmental
hazards would result depending on which program was
invoked. Under the latter Act, the emphasis is upon
exploitation of oil, gas and other minerals, with, to
be sure, all necessary protective controls. Under the
Sanctuaries Act, the prime management objectives are
conservation, recreation, or ecological or esthetic

values.   Drilling and mining may be allowed, but the primary emphasis remains upon the other objects.   The marked differences in priorities could lead to different administrative decisions as to whether particular parcels are suitable for oil and gas operations.   And should there be particular areas of Georges Bank that are uniquely important to the fishery, for example, a key breeding area or the like, the management by the Secretary of Commerce, the administrator of the Fishery [Conservation and Management] Act, rather than by the Secretary of the Interior might be advantageous. At least the question seems worth exploring. [*Massachusetts* v. *Andrus*, 1979a]

The department decided, as a result of this language that it should prepare a supplementary EIS to consider the marine sanctuary alternative, and to incorporate in the analysis both the marine sanctuary option and the recent changes in the OCS Lands Act.   The department hoped by doing so to avoid further litigation on these issues, but three months after the appeals court decision the Conservation Law Foundation took up the court's suggestion that management of Georges Bank should be by the Department of Commerce rather than the Department of the Interior and formally nominated Georges Bank as a marine sanctuary.

On May 10, 1979 the CLF submitted to Secretary of Commerce Juanita Kreps a nomination of the entire Georges Bank area as a marine sanctuary (Conservation Law Foundation, 1979).   CLF's nomination was on behalf of a number of fishing groups in Massachusetts.   The basic thrust of the nomination was to transfer management of Georges Bank to the Department of Commerce.   Georges Bank would be declared a sanctuary whose priority use would be the production of food.   Protection of the fisheries resources was to be the standard against which all other uses, including oil and gas operations, were to be judged.   The arbiter of all competing uses was to be the New England Fisheries Management Council, a group of state and fishing industry representatives created to be the principal decision maker regarding the management of fish stocks by the Fisheries Conservation and Management Act of 1976.   The nomination thus sought to explicitly implement the Court of Appeals' suggestions regarding the different management objectives of the OCS Lands Act and the Marine Sanctuaries Act.

In the nomination, CLF advanced three principal arguments for the creation of a Georges Bank Marine Sanctuary.

1. A marine sanctuary is inherently more flexible and allows greater protection of marine resources, since the secretary of commerce is authorized to establish whatever regulations are necessary in establishing a marine sanctuary.
2. The OCS Lands Act, even as amended, requires a balance between oil and gas development and other ocean uses. But simply balancing oil and gas development against fisheries is not sufficient for Georges Bank, because of the unique productivity of the region.
3. The Department of the Interior is too committed to oil and gas development to provide the protection required for Georges Bank.

At the same time that the marine sanctuary nomination was being submitted, the Bureau of Land Management published its Draft Supplemental EIS (DSEIS), and in June the hearings on the new statement were held. The focus of the hearings was the question of marine sanctuaries, although debate continued on the general question of leasing on Georges Bank.

For the most part, the New England states, all of which had undergone changes in administration in 1979,* were cautious about the marine sanctuary nomination. They recognized that the essence of any marine sanctuary was contained in the regulations of the sanctuary, and that the Georges Bank nomination had not progressed far enough to make a firm decision. "It is difficult for the benefits of a Marine Sanctuaries authority, since one must speculate as to the content of the future regulations in the proposed Georges Bank Marine Sanctuary . . ." (Massachusetts, 1979a). Since NOAA would have to go through a lengthy development and hearings process before a Marine Sanctuary could be designated, Maine, New Hampshire, and Massachusetts all agreed that careful consideration of the Marine Sanctuary option was

---

*The four changes in administration were as follows: Maine: Joseph E. Brennan replaced independent James B. Longley, who retired; Massachusetts: Edward J. King replaced Michael Dukakis, whom he had defeated in the Democratic Primary; New Hampshire: Democrat Hugh J. Gallen defeated Republican incumbent Meldrim Thomson; Rhode Island: Democrat J. Joseph Garrahy replaced Republican incumbent Philip Noel.

required, but that the Lease Sale could go ahead on schedule
(Massachusetts, 1979a; Maine, 1979; New Hampshire, 1979).
The oil industry representatives who appeared at the public hearing spent the majority of the time arguing against
the whole idea of a Marine Sanctuary.

> API's [the American Petroleum Institute] CZM Steering
> Committee is firmly of the opinion that a marine sanc-
> tuary designation the magnitude of the CLF's Georges
> Bank nomination is unnecessary, and that adequate reg-
> ulations exist to thoroughly protect the resource
> values of the Bank while still permitting the develop-
> ment of highly needed oil and gas energy resources for
> the region and the nation. [American Petroleum Insti-
> tute, 1979]

For the environmental groups at the hearing, including
the CLF, the Sierra Club, and the Natural Resources Defense
Council the desirability of a marine sanctuary was self-
evident, and their principal argument was that the lease
sale should be delayed pending the completion of the sanctuary designation process by NOAA.

> We have been assured many times that there are ade-
> quate safeguards in place to proceed with OCS leasing
> at this time. Yet we are told in this Draft Supple-
> ment that routine pollutants will be discharged, drill
> cuttings and muds are toxic but their effects are un-
> known, [and] regulations to implement the OCS Lands
> Act have not been implemented. This Draft Supplement
> presents some compelling arguments in favor of the
> delay to the lease sale. Considering the risks we see
> no other alternative. [Rockefeller, 1979]

In July, 1979 the final act of the Georges Bank Lease
Sale story began. The Department of Commerce held several
public hearings in New England to receive comment on an
issue paper concerning the CLF nomination. These hearings
were largely replays of the DSEIS hearing. Also in July the
Department of the Interior issued its second Proposed Notice
of Sale for Lease Sale 42, and requested governors' comments. The lease sale was scheduled for October 30, 1979.
In August, the Final Supplemental EIS was published,
which included a discussion of the CLF nomination. But by
this time the action had shifted to Washington, where the
Carter administration was faced with a major decision.
Carter and Andrus had lobbied very hard for the OCS Lands

Act, arguing that it was essential for future harmonious development of the OCS. The support of the administration was considered essential in finally getting the amendments past Congress. But having lent the full weight of their support to the amendments as the key to the management of the OCS, the CLF nomination asked, in essence, that for Georges Bank the new regime be tossed out and replaced with an entirely different one. For the Carter administration, which had long since declared the energy issue the "moral equivalent of war," changing the process for Georges Bank would only fuel support to restrict OCS operations in other frontier areas.

So in late September, the Department of the Interior and the Department of Commerce announced a Management Plan for Georges Bank. The essential feature of the plan was the withdrawal of Georges Bank from consideration as a marine sanctuary and the creation of a Biological Task Force. This task force (which came to be know as the BTF) was to be composed of the U.S. Geological Survey (USGS), the Bureau of Land Management, the Fish and Wildlife Service, the National Marine Fisheries Service, and the Environmental Protection Agency. The states and the New England Fisheries Management Council* were to be nonvoting members. The BTF was created by a formal charter, which was an interagency agreement signed by the heads of all five federal agencies involved. The principal function of the task force was to assist in implementing Stipulation 2 by recommending to the USGS district supervisor which areas of Georges Bank were to be considered environmentally sensitive, the measures which were to be taken to protect these areas, and any monitoring studies required to analyze the effects of oil and gas operations on the biological resources of Georges Bank.

The charter of the task force set forth an elaborate set of procedures for dealing with disputes that might arise during the task force's work. Since the task force could only recommend actions to the district supervisor, an appeals procedure was established whereby any two of the members could appeal to the director of the geological survey

---

*The New England Fishery Management Council was one of the regional councils established by the Fishery Conservation and Management Act of 1976 to oversee regulation of the in the 200-fishery-mile zone.

and ultimately to the Secretary of the Interior in the event of disputes. The fear was that with three of the task force members representing the Department of the Interior agencies, the other two agencies would be consistently outvoted on important issues.

Two weeks after the announcement of the Georges Bank Management Plan, the notice of sale for Lease Sale 42 was published in the federal register. The Carter administration decision on the marine sanctuary was greeted with dismay in New England. States, such as Maine and Massachusetts, who had argued that the sanctuary designation process should proceed while the lease sale went ahead, were upset by the abrupt withdrawal of the sanctuary. The composition and limited authority of the Biological Task Force were regarded as inadequate (Brennan, 1979). Thus it is not surprising that with the second notice of sale came the second Georges Bank lawsuit.

The plaintiffs in Georges Bank II again included the Conservation Law Foundation, the Commonwealth of Massachusetts, and the state of Maine. Maine had sought admission to the 1978 lawsuit, but had not been admitted by the Court. The arguments in the lawsuit were generally similar to those made in the first case: inadequate Environmental Impact Statement, inadequate consideration of endangered species, and "arbitrary and capricious" decision making in the decision to withdraw the marine sanctuary from consideration. The argument that the Secretary of the Interior had a duty to protect the fisheries of Georges Bank was again advanced, but since the court of appeals had ruled that such a duty did not flow from the OCS Lands Act, the plaintiffs in the case argued that the Secretary's duties stemmed from the common law (Massachusetts, 1979b; Conservation Law Foundation, 1979b).

A typographical error in the original federal register version of the notice of sale had to be corrected and so the sale was postponed from October 31 to November 5. It was thus on October 31 that oral arguments began in Boston on the request for a preliminary injunction against the sale. The case was heard by District Judge John McNaught rather than Judge Arthur Garrity, and much of the argument repeated the earlier version of the lawsuit.

Two days before the lease sale was to be held, the Court ruled against Massachusetts and CLF (*Massachusetts* v. *Andrus*, 1979b). McNaught had not been convinced that either the Department of the Interior or the Department of Commerce had violated any of the laws cited, and so denied the request for a preliminary injunction on the grounds that plaintiffs were unlikely to succeed when a trial on the merits of the case was held.

Massachusetts and CLF immediately appealed to the First Circuit Court of Appeals, but late on November 4 the appeals court also refused to grant the preliminary injunction. The appeals court did stay its decision to allow an appeal to the Supreme Court. The next morning, as oil company officials were gathering in a hotel in Providence for the opening of bids, two lawyers from CLF and Massachusetts flew to Washington to make their appeal to Supreme Court Justice William Brennan. About two o'clock in the afternoon of the fifth, Brennan issued a stay of the lease sale pending review by the full Court. The Department of the Interior was forced once again to cancel the lease sale, and the oil companies headed out of Providence bearing their unopened bids.

On Friday, November 8, the full Supreme Court vacated the stay that Brennan had issued. Ten days later, the Department of the Interior issued the third notice of sale for Lease Sale 42, this time for December 18, 1979, again in Providence. At this point the status of the lawsuit was that a trial on the merits of Massachusetts and CLF's request for a permanent injunction was still to be held. A full argument before the court of appeals was also pending. Once the third notice had been issued, the appeals court arguments were made, but on December 17, the court again refused to overturn McNaught's decision.

The morning of December 18 was a replay of November 5. The Massachusetts and CLF lawyers were again on a plane bound for Washington as the oil company people again gathered in Providence. At 10:30 in the morning Brennan again issued a stay, but at 11:30 he reversed himself and refused to issue a stay. One hour later, Frank Basile, the manager of the New York OCS office opened the first bid, and Lease Sale 42 was finally underway.

At the end of the day, 73 of the 116 tracts offered had received bids. A total of $1.27 billion had been exposed in the bidding, with a high bid total of $828 million. The largest bid for a single tract was $80.7 million by a group lead by Mobil. Later, 63 bids would be accepted by the Bureau of Land Management. It was four years from the time that Lease Sale 42 had originally been scheduled.

Now, having won their leases the oil companies still had to go through the process of getting the required permits. In both lawsuits the federal government and the oil companies who joined the suit on the side of the federal government had argued that the lease sale was not an irrevocable decision since there was ample opportunity for the Department to deny permits or halt drilling later in the process. The specter of this possibility was in the back of everyone's mind as the leases were signed and the permitting process begun. No one was betting that the next step in the process would be any easier than the previous ones. The view of most was summed up by the newly appointed North Atlantic District Supervisor who said drilling on Georges Bank would begin when the Philadelphia Phillies won the World Series. At the time, the Phillies had not won a World Series in over 60 years (Dannenberger, 1980).

The permitting process focused on three major arenas: The federal agencies, the states, and the Biological Task Force. Three federal agencies are required to issue permits for exploratory drilling. The U.S. Geological Survey has overall regulatory responsibility for all aspects of OCS drilling. The oil company must submit an exploration plan and environmental report to the USGS for approval. The plan and report were required by the 1978 OCS Lands Act Amendments and are meant to be detailed descriptions of the proposed drilling, the equipment to be used, and the likely impacts.

The Environmental Protection Agency issues a National Pollution Discharge Elimination System (NPDES) permit under the Clean Water Act. This permit covers all routine discharges from the drilling rigs, most importantly the discharge of drilling muds, cuttings, and oil contaminated deck drainage.

The Army Corps of Engineers is responsible for issuing a permit for any structure to be placed in the navigable waters of the United States. The corps reviews the proposal for any impacts on national defense activities or navigation.

The states are granted substantial review authority when it comes to these permits. That authority stems from the 1972 Coastal Zone Management Act, which allowed states with federally approved Coastal Zone Management Programs to review activities that require a federal permit and which "affect land or water uses in the coastal zone" for consistency with those approved programs. Since federal agencies were forbidden to issue permits until the states had reviewed the proposed drilling, the state became key actors in this stage. Four of the New England coastal states had Coastal Zone Management Programs: Connecticut, Rhode Island, Massachusetts, and Maine. Connecticut, because of its location on Long Island Sound, did not actively participate in reviewing the Georges Bank plans, and Rhode Island's governor was firmly committed to supporting the drilling, so Rhode Island's review was perfunctory. This left Massachusetts and Maine as the major participants in the federal consistency reviews, and it was not lost on anyone that these were the states that had just sued to halt the lease sale.

The final arena was the Biological Task Force (BTF), the interagency group whose creation had caused so much controversy and anger the preceeding fall. Most of 1980 was spent by the oil companies in conducting the shallow geophysical surveys on their tracts and preparing their exploration plans, environmental reports, and permit applications. Thus the BTF held center stage during the majority of this period. The task force held its first meeting at the USGS North Atlantic District Office in Hyannis, Massachusetts about two weeks after the leases had been signed. The NOAA representative was elected presiding officer and the task force immediately defined its principal task as designing a research and monitoring program to determine what effects drilling-mud discharges would have on Georges Bank. A monitoring subcommittee was formed and charged with designing a monitoring program. The EPA representative took the lead in the subcommittee and the program design, which somewhat assuaged the fears of those who thought the task force would be dominated by the Department of the Interior representatives.

The monitoring subcommittee and the full task force held a number of meetings throughout the late winter and spring, and went through a number of drafts and discussions on the content of a monitoring program. Finally, in July, the task force approved a monitoring program that consisted primarily of a system of 40 monitoring stations set in a radial pattern around one of the rigs. The stations were to be sampled for changes in species diversity, sediment and water column chemistry, and heavy metals accumulation. The program concentrated on the short-term impacts of drilling discharges, and described long-range monitoring problems only in general terms. The task force unanimously approved the program in July, and forwarded it to the district supervisor for consideration.

Despite the unanimous support of the federal agencies for the program, the states had been skeptical of the program from its early stages. The states criticized the program as lacking any ability to generate information useful to decision makers, as well as any systematic hypotheses concerning the fate and effects of drilling fluids and too much emphasis on field studies in the absence of any solid evidence on the precise toxic nature of the fluids.

Although the charter assigned the decision-making authority on the task force's recommendations to the district supervisor, the first recommendation of the BTF was for a
multi-year, multi-million-dollar research program. The
issue was quickly passed on to higher levels of the Department of the Interior in Washington.

The department initiated a review of the program, first
by its own staff, and second by two of its advisory committees. In August, one of those groups composed of state,
federal, and private citizen members from the North Atlantic
area (North Atlantic Technical Working Group, 1980) recommended that the department reject the task force's proposal
and require a rewrite. Two months later, the Scientific
Advisory Committee, composed of marine scientists from
around the country, concurred in the recommendation that the
Department of the Interior reject the BTF monitoring program, citing the same shortcomings that the states had com-
plained of during the program's development. In November
the department accepted these recommendations and sent the
program back to the BTF for redesign.

The task force went back to the drawing board in December, and in January a new version of the program was ready
for review. The new program met most of the objections that
had originally been raised. The sampling program for the
rig study was redesigned to intensively study areas where
drilling muds were projected to go on the basis of known
characteristics of the Georges Bank current system. Expensive water-column studies were deleted in favor of concentration on the ocean-bottom environment, where the muds were
expected to accumulate. And a long-term monitoring effort
to examine the possibility of transport of drilling fluids
far from the drilling sites was established. In March, 1981
the department approved the plan and agreed to provide over
$1 million for the first year of the effort.

By this time, however, attention was focused back on the
states. The first exploration plans were submitted for review in December 1980. The plans from Getty, Mobil, and
Exxon were approved by the USGS within 30 days, as required
by the OCS Lands Act. That left only Maine, Massachusetts,
and the EPA to act.

In the winter and spring of 1981 the states held public
hearings on their consistency reviews of the exploration
plans, and carried on extensive correspondence with the
three oil companies in an attempt to understand the complex-
ities of the drilling equipment and procedures, and to evaluate the several hundred pages of environmental information
submitted with each plan. The states spent considerable

time as well with EPA's Region I office in Boston discussing the contents of the NPDES permit, since the question of drilling mud discharges had come to dominate the concerns of most, just as it had dominated the BTF.

In May, 1981 EPA issued its Draft NPDES permit for public review. In late May EPA, the USGS, Maine, and Massachusetts held the last of their meetings, and agreed upon the basic conditions that would be required for the discharges: The companies would be required to use only a limited number of drilling muds whose chemical characteristics and toxicity had already been established. Detailed records of the mud composition, with periodic laboratory tests to check the muds' chemical composition and toxicity would be required. The discharge of drilling muds would also be limited to a maximum of 30 barrels per hour; the discharge had to take place ten meters below the surface to avoid the plankton which concentrated at the surface, and would have to be diluted prior to discharge at a ratio of ten parts seawater to one part whole mud.

These conditions were included in the EPA permit, and in the consistency approvals that Maine and Massachusetts issued in June. The oil companies, which had opposed the extensive mud testing and discharge requirements, did not file an appeal against the EPA permit as they had done in other frontier areas, in large part because of the agreement between state and federal agencies.

At the end of June, 1981 all the federal permits were issued with concurrence from Maine and Massachusetts, and Shell and Exxon began moving rigs toward Georges Bank. The Shell rig left the Gulf of Mexico at the beginning of July, and the Exxon rig left the Baltimore Canyon area off New Jersey a few days later. On July 9, the first research cruise of the BTF monitoring program left Woods Hole, Massachusetts to conduct the predrilling surveys called for by the program. Two weeks later the first wells were begun. It was not known whether this was the beginning of an exploitation of Georges Bank in a new and radically different way that would continue for years, perhaps decades. But July 24, 1981 was clearly the end of the beginning.

THE SEARCH FOR A NATIONAL OCEAN POLICY

The period from the aftermath of the Arab oil embargo to the opening of the first exploratory oil wells on Georges Bank was the period in which the United States spent the most time and effort debating the issues involved in using ocean

resources. The most fundamental questions about ocean resource management were raised, and through the process of
getting drilling started in New England, these questions
were at least partly answered. Three issues lay at the
heart of the debate.

1. Who should be the resource manager?
2. How much information is needed to make management
decisions?
3. What are the tradeoffs when new resources are to
be developed?

Who should be the Resource Manager?

When Rogers Morton announced the Department of the Interior's accelerated leasing schedule in November, 1974, the
federal government did not even have clear title to the
Atlantic OCS. The lawsuit against the State of Maine for
attempting to sell offshore leases was still pending before
the Supreme Court, with no immediate settlement forseeable.
But even after the Court finally awarded title to the federal government, the issue of what the respective roles of the
federal and state governments were to be remained unsettled.

The state governments, beginning with the 1975 push for
amendments to the OCS Lands Act, insisted that they play a
full role in the management of OCS oil and gas; at that time
Maine and Massachusetts spoke of a guaranteed partnership
with the federal government. The reason for the states'
insistence on a role was the perception that the risks and
benefits from offshore oil development would be unevenly
distributed. If oil and gas were found, the energy supplies
would add to the national resources, but, particularly in
the case of oil, the New England region would see almost no
direct benefit. Oil would have to be transported to the refineries in the mid-Atlantic states, where it would be fed
into the northeastern markets. Some oil would get to New
England, but the precise amount could never be predicted.
If gas were found, it would have to be pipelined ashore
where it would be processed and fed into existing distribution networks. More gas than oil would stay in New England,
but these benefits were considered highly speculative.

On the other hand, from the perspective of Massachusetts
particularly, a major oil spill on Georges Bank had the potential for large-scale disruptions of the fishing industry.
There was also fear that an oil spill could severely damage
the tourist industry in such highly used areas as Martha's

Vineyard, Nantucket, and Cape Cod. Maine, with the second-largest fishing industry in New England, was equally concerned about the possibilities of an oil spill.

The importance of the tourist and fishing industries in the economies of the coastal areas of New England was striking. In areas such as New Bedford, Massachusetts, fishing was almost the only major industry. This was also the period when New England was leading the effort to establish a 200 mile fisheries management zone, due primarily to the heavy overfishing of Georges Bank by foreign fishing fleets in the late 1960s and early 1970s. There was thus great hope for the future of the New England fishing industry. Tourism was also a mainstay of coastal areas, particularly in such areas as Nantucket and Cape Code, which were also the closest coastal areas to the lease sale area.

Governors could hardly ignore the problems that might be caused if major disruptions occurred in the economic base of substantial portions of their states. Attention to these potential problems was heightened, at the same time, by the fact that all the New England states were undergoing development of their Coastal Zone Management programs during this period. CZM program development forced the states to increase awareness of the value of their coastal resources, and this increased attention carried over to concern about coastal and ocean resource management questions in general and to offshore oil in particular.

The CZM program was one which the states voluntarily participated in, with two major carrots set before them: First, the promise of substantial federal funds to implement the management programs once they had been approved. Second, the authority to review federally conducted and permitted activities for consistency with the state's approved program. As it turned out, the federal consistency review was far more important to the states in dealing with the actual drilling on Georges Bank. The 30 day period allowed the governors by the OCS Lands Act was used, but the focus of state efforts was clearly on consistency.

In the years before the OCS Lands Act Amendments, the states were not sure that their programs would in fact be approved by the federal government, or even that the necessary compromises among coastal interests in the states could be achieved so that an acceptable CZM program could be developed. In addition, there were still serious uncertainties surrounding the entire consistency process. It was not clear whether OCS activity fit in to the consistency review. It took a specific amendment to the CZMA in 1976 to make it explicit that exploration and development plans would be

included under the section authorizing the review of federally licensed and permitted activities. The section of
the CZMA that authorized review of federal activities directly affecting the coastal zone was also an unknown
quantity. As late as 1981 the meaning of this provision was
still disputed.

Added to their perceptions of the benefits and risks involved and the importance of their coasts, there was also a
fair portion of distrust of the federal government. The
causes of this distrust, other than the background level
that exists between all levels of government in the federal
system, were primarily three. First, the Ford administration seemed to want to go ahead with its leasing plans regardless of states' expressed opinions. The opposition of
the administration to the OCS Lands Act Amendments was the
most visible cause of state distrust during 1975-76.

Under Jimmy Carter, the administration was no longer
opposed to an expanded role for the states and the other
provisions of the OCS Lands Act Amendments. The refusal to
implement the oil spill and fisherman's contingency funds
administratively prior to Lease Sale 42 was frustrating,
particularly for Massachusetts that viewed these as absolute
prerequisites to oil and gas exploration and that believed
that the department could have established these measures
administratively, became the second major source of distrust.

Third, there was the view, which was by no means confined
to the states, that the Department of the Interior was
simply too pro-oil. This feeling was most clearly expressed
by CLF in the marine sanctuary nomination, which was nothing
less than an attempt to take OCS development away from the
Department of the Interior and give it to the Department of
Commerce, which had responsibilities for fisheries conservation and development.

How Much Information Was Necessary to Make Decisions?

There was substantial concern expressed throughout the
Georges Bank leasing process about how much was known, how
much was not known, and what had to be known in order to
make decisions about oil and gas development. There was
fear that since relatively little was known about the
Georges Bank ecosystem, including such things as the current
patterns and the location of fish spawning and nursery
areas, that OCS development would take place in ignorance of
the potential impacts.

A great deal was made during the review of the EIS of the experience in the Gulf of Mexico, where drilling had been going on for more than 30 years without any documented impacts. But the example did not convince many, who pointed out that the environment of the Gulf of Mexico was different from that of the North Atlantic, and that no baseline studies had ever been done to identify whatever impacts might have occurred.

There was also concern that not enough was known about the impacts of oil-generated pollutants such as drilling muds, cuttings, and formation waters. Concern about drill muds was not great in the early stages of the leasing process, but became a central concern during the supplemental EIS review, and in the period after the lease sale. The composition of drill muds, their fate and effects when released into the marine environment, and the toxicity to marine organisms were all intensely debated. During the development of the BTF monitoring program, there was criticism of all the previous investigations of drilling-mud discharges that pointed out that a great deal of poor science had been done in those investigations (Spiller and Rieser, 1981).

There were thus basic questions about the nature and characteristics of Georges Bank, about the offshore oil and gas industry's effects, and about the research into those effects. For many, the combination was sufficient to argue that leasing should be postponed until a great deal more information was gathered.

What Are the Tradeoffs?

Although the role of the states and the question of information were important, the heart of the Georges Bank controversy was clearly the question of tradeoffs. For many in New England, the tradeoff questions were starkly clear. The coming of oil and gas to Georges Bank was believed by all to inevitably doom the fishery, and it simply made no sense to try to acquire two weeks worth of oil for the United States if it meant giving up a fishery that had the potential to feed people virtually forever.

Two conclusions were drawn from those who emphasized that oil and gas development on Georges Bank would involve trading off fish for fuel. For some, oil and gas development would never be acceptable under any circumstances. For others, there was a recognition that oil and gas development was probably inevitable, and should only be done with priority given in all decisions to maintaining the fishery.

This latter was the Conservation Law Foundation's position, as expressed in two major ways. In the two court suits CLF tried to force the interpretation that the Secretary of the Interior was responsible for protection of the fisheries just as much as he was responsible for oil and gas development. When the appeals court ruled in round one of the lawsuit that CLF and the district court had misinterpreted the OCS Lands Act, CLF was back arguing the Secretary of the Interior's duties to the fisheries as a result of the common law.

This view was also at the center of the marine sanctuary nomination. The primacy of the fishery resource was the standard against which all activity was to be judged in the Georges Bank Marine Sanctuary. The importance of the fishery resources of Georges Bank was so great that no balance between fish and oil could be acceptable.

The concept of a balance was the center of the other view of the tradeoffs involved. The Appeals Court noted in its 1979 decision that Congress in the OCS Lands Act required the secretary to balance the energy needs of the country against the other resource needs. The legislation implicitly stated the view that a balance was possible and left it up to the secretary to strike the balance.

Although there were serious disagreements about timing and other issues, the federal government (under Carter) and the state governments were in agreement that there were certain measures that could be taken to minimize the tradeoffs. A requirement that the best available and safest technologies be required of all operators was one such measure. The oil spill and fishermen's contingency funds that would lessen the financial risk to the fishing and tourist industries in the event of damage was another. There was also a requirement that all oil personnel undergo a training program to familiarize them with the characteristics of the New England fishing industry. The Department of the Interior also emphasized the broad range of regulatory authority over oil and gas operations it possessed, as well as the specific requirements of Stipulation 2 concerning areas of biological significance.

The history of the Georges Bank Lease Sale shows that the principle of balanced development was the one that was finally adopted. The victory of this principle was clouded by the controversy surrounding the withdrawal of the marine sanctuary nomination from consideration by the federal government, but the refusal of the courts to overturn that decision made it unlikely that OCS management would be based on the primacy of the fishery resource, at least for Georges Bank as a whole.

The Georges Bank Lease Sale also shows how the other two fundamental questions were answered. The resource manager of the OCS is clearly the federal government, but the states have substantial input to the process and, through the consistency review, a guaranteed voice in federal-permitting decisions. In the end, the federal-state cooperation was proven successful during the post-sale permitting process, although the relative importance of the Coastal Zone Management and OCS Lands Acts was the reverse of what might have been expected in 1975 or 1976.

Finally, although there were no definitive answers to any of the questions about the nature of the Georges Bank environment or about the impacts of oil operations, the Biological Task Force Monitoring Program, coupled with the EPA permits provides an approach that will serve as an example of OCS environmental management under conditions of uncertainty. The restrictions on the depth, rate, and dilution of the drill-mud discharges were designed to minimize the exposure of the benthic- and water-column biota that were considered to be the most sensitive. At the same time an intensive study of both long-term and short-term impacts would be conducted to be certain that impacts were being avoided. The use of monitoring is expected to become a major focus of OCS environmental studies in the future. And so the question raised at the beginning recurs: Why did it take seven years for drilling to begin on Georges Bank?

It should be clear that the delays were not the result of ill-informed or unreasoning opposition to the oil companies or oil drilling. During the course of the lease sale planning there were three major oil spills: The tanker *Argo Merchant* ran aground on Nantucket shoals in December, 1976 and spilled 50,000 barrels of oil onto Georges Bank just a few days before the DEIS public hearings. In 1978, a large blowout and spill occurred in the North Sea in Phillips Petroleum's Ekofisk field, and in 1979 the blowout in the Baia de Campeche off the Yucatan peninsula became the world's largest oil spill by several orders of magnitude. All these events reinforced the views of those who believed oil and gas development was inherently unsuited to Georges Bank.

The major opponents of the lease sale, Massachusetts and the Conservation Law Foundation, were not opposed to drilling itself. They sought, along with Maine and other New Englanders, to find ways to minimize risk, to improve management, and to provide some recognition of the importance of the competing coastal and ocean resources.

In the period from the announcement of the sale to the first cancellation of the sale following the first court suit, it was the perception of the uneven distribution of risks and benefits and the inability of the existing federal policies to deal with this distribution that motivated concern. Early worries about onshore impacts were largely relieved by the passage of the Coastal Energy Impact Program in 1976, but the OCS Lands Act amendments were the major objective. It was the lack of these amendments that the major reason for the injunction that halted the lease sale in 1978.

But after the amendments were passed, the issue was still not resolved. The marine sanctuary was presented almost immediately as an alternative, even a repudiation, of the management principles embodied in those amendments. There was an element of chance in CLF's ability to raise the marine sanctuary as an alternative. The possibility of a marine sanctuary was not one that had been seriously raised during the DEIS hearings or the earlier discussions. In the first court suit CLF had raised the issue as an allegation of a violation of the National Environmental Policy Act requirement that alternatives to a proposal be considered in an EIS. The district court had upheld the charge, but did not present it as a major reason for issuing the preliminary injunction. But when the appeals court pointed out the implications of the different management objectives of the Marine Sanctuaries and OCS Lands Acts, the possibility of using the Department of Commerce's broad powers in a marine sanctuary was simply too attractive to pass up.

Despite this element of legal serendipity, however, the issue in the marine sanctuary nomination was still the basic management structure and principles. This was the heart of the Georges Bank oil story. Despite the clear national energy problems, there was simply not sufficient consensus on management policies for the oceans to allow a rapid development of OCS resources. The Georges Bank Lease Sale was caught up in, and a prime motivating force for, the development of a national ocean management policy, and it was not until there were at least broad answers to the fundamental management questions that drilling could begin.

There are legitimate questions about the reasons for the length of time it took to arrive at a consensus and the broad outlines of a policy. Whether exploration could have begun as a result of the first lease sale if the OCS Lands Act Amendments had been passed any time during 1975, 1976, or 1977 is certainly an open question, but it will take a detailed study of the history of those amendments to get close to an answer.

The ultimate judgment on the wisdom of the policy and its application to Georges Bank lies many years in the future. Too many variables will affect the outcome, some of which, such as the future U.S. energy needs and occurrence of any oil spill are not a direct function of the policy itself.

The decade of the 1980s will start out with comprehensive and detailed federal and state programs and policies in place for the management of coastal and ocean resources. The 1970s provided for a system of oil and gas development based on federal-state cooperation, a new commitment to assessing the impacts of oil operations on the environment, and a basic principle of recognizing that oil and gas development must be balanced against the need to conserve and develop other ocean resources.

The story has not ended of course, either for Georges Bank or the nation as a whole. If a find is made on Georges Bank, a whole new round of decisions will be required for development and production. There are three more lease sales scheduled through 1986, and probably more after that. And future exploration is expected to move onto the continental slope, in waters up to 10,000 feet deep.

At the same time, the federal government under a new administration is attempting to change many of the fundamental management practices regarding OCS leasing. But the basic policy structure remains, and the lesson of Georges Bank is clear. The OCS can only be developed when there is a consensus that the risks will not be unevenly distributed, that all interests will be considered, and the decision making will be made on the basis of sound scientific information. The failure to understand the need for that consensus was the prime cause of the delay in Georges Bank drilling.

REFERENCES

Allen, R.B. 1975. Testimony at New London, Conn. before the Ad Hoc Select Committee on the Outer Continental Shelf on the Outer Continental Shelf Lands Act Amendments of 1975 (H.R. 6218), p. 1936.

American Petroleum Institute. 1979. Comments on Draft Supplemental Environmental Impact Statement in B.L.M., Final Supplemental Environmental Impact Statement for Lease Sale 42 (July, 1979), p. 378.

Baldwin, P.L., and Baldwin, M.F. 1975. *Onshore Planning*

for Offshore Oil: Lessons from Scotland. Washington, D.C. The Conservation Foundation.

Brennan, Joseph E. 1979. Letter to the President, September 28, 1979.

Bureau of Land Management. 1976. *Draft Environmental Impact Statement for Lease Sale 42.*

Conservation Law Foundation of New England. 1977. Comments on Draft Environmental Impact Statement in Bureau of Land Management, Final Environmental Impact Statement for Lease Sale 42, p. 1559.

Conservation Law Foundation of New England. 1979a. Letter and Nomination Paper concerning a Georges Bank Marine Sancturary submitted to Secretary of Commerce Juanita Kreps, May 10, 1979.

Conservation Law Foundation of New England. 1979b. Brief in Support of Motion for Preliminary Injunction, *Conservation Law Foundation of New England v. Andrus,* First Circuit Court of Appeals, Boston, Mass., Case number 78-0186-MC, October 1979.

Dannenberger, E.P. 1980. Personal communication.

Dukakis, Michael J. 1975a. Testimony at Boston, Mass. before the Ad Hoc Select Committee on the Outer Continental Shelf on the Outer Continental Shelf Lands Act Amendments of 1975 (HR 6318), p. 2032.

Dukakis, Michael J. 1975b. Statement submitted to the Record at Boston, Mass. to the Ad Hoc Select Committee on the Outer Continental Shelf on the Outer Continental Shelf Lands Act Amendments of 1975 (HR 6318), p. 2035.

Frazier, Ronald F. 1975. Letter to James A. Burns, M.C. included in the record of the Ad Hoc Select Committee on the Outer Continental Shelf on the Outer Continental Shelf Lands Act Amendments of 1975 (HR 6318), p. 2254.

Gifford, P. 1975. Testimony at Boston, Mass. before the Ad Hoc Select Committee on the Outer Continental Shelf on the Outer Shelf Lands Act Amendments of 1975 (HR 6318), p. 2736.

Kendall, Richard E. 1977. Statement at Hearing on Draft Environmental Impact Statement in Bureau of Land Management, Final Environmental Impact Statement for Lease Sale 42, p. 1499.

Longley, James B. 1975a. Testimony at Boston, Mass. before the Ad Hoc Select Committee on the Outer Continental Shelf on the Outer Continental Shelf Lands Act Amendments of 1975 (HR 6318), p. 2078.

Longley, James B. 1975b. Statement submitted to the Record at Boston, Mass. to the Ad Hoc Select Committee on the Outer Continental Shelf on the Outer Continental Shelf

Lands Act Amendments of 1975 (HR 6318), p. 2035.

McAuliffe, Claton P. 1977. Statement at Hearing on Draft Environmental Impact Statement in Bureau of Land Management, Final Environmental Impact Statement for Lease Sale 42, p. 1538.

Maine, State of, State Planning Office. 1977. Comments on Draft Environmental Impact Statement in Bureau of Land Management, Final Environmental Impact Statement for Lease Sale 42, p. 1520.

Maine, State of, State Planning Office. 1979. Comments on Draft Supplemental Impact Statement in Bureau of Land Management, Final Supplemental Environmental Impact Statement for Lease Sale 42 (July, 1979), p. 347.

Massachusetts, Commonwealth of, Office of Coastal Zone Management. 1977. Comments on Draft Environmental Impact Statement in Bureau of Land Management, Final Environmental Impact Statement for Lease Sale 42, p. 1460.

Massachusetts, Commonwealth of, Executive Office of Environmental Affairs. 1979a. Comments on Draft Supplemental Environmental Impact Statement in Bureau of Land Management, Final Supplemental Environmental Impact Statement for Lease Sale 42 (July, 1979), p. 350.

Massachusetts, Commonwealth of. 1979b. Brief in Support of Motion for Preliminary Injunction, *Massachusetts v. Andrus*, First Circuit Court of Appeals, Boston, Mass., Case number 78-0186-MC, October 1979.

Massachusetts v. Andrus, Opinion of the Court (1978).

Massachusetts v. Andrus, 594 Federal Reporter Second Series (1979a), p. 872.

Massachusetts v. Andrus, 481 Federal Supplement (1979b), p. 685.

Matthews, Charles D. 1977. Statement at Hearing on Draft Environmental Impact Statement in Bureau of Land Management, Final Environmental Impact Statement for Lease Sale 42, p. 1557.

New Hampshire, State of: State Planning Office. 1979. Comments on Draft Supplemental Environmental Impact Statement in Bureau of Land Management, Final Supplemental Environmental Impact Statement for Lease Sale 42 (July, 1979), p. 349.

North Atlantic Technical Working Group. 1980. Resolution adopted at meeting of August 18, 1980.

Pines, Lois G. 1977. Statement at Hearing on Draft Environmental Impact Statement in Bureau of Land Management, Final Environmental Impact Statement for Lease Sale 42, p. 1499.

Rockefeller, Marsha. 1979. Comments on Draft Supplemental

Environmental Impact Statement in Bureau of Land Management, Final Supplemental Environmental Impact Statement for Lease Sale 42 (July, 1979), p. 399.

Sierra Club. 1977. New England Chapter Statement at Hearing on Draft Environmental Impact Statement in Bureau of Land Management, Final Environmental Impact Statement for Lease Sale 42, p. 1534.

Sloan, Lucy. 1975. Testimony at Boston, Mass. before the Ad Hoc Select Committee on the Outer Continental Shelf on the Outer Continental Shelf Lands Act Amendments of 1975 (HR 6318), p. 2736.

Spiller, J. and Rieser, A. 1981. *Regulating Drilling Effluents on Georges Bank and Mid-Atlantic Outer Continental Shelf: A Scientific and Legal Analysis.* Boston, Mass.: New England River Basins Commission.

Studds, Gerry E. 1977. Letter to Bureau of Land Management on Draft Environmental Impact Statement in Bureau of Land Management, Final Environmental Impact Statement for Lease Sale 42, p. 1458.

Thomson, Meldrim L. 1975. Testimony at Boston, Mass. before the Ad Hoc Select Committee on the Outer Continental Shelf on the Outer Continental Shelf Lands Act Amendments of 1975 (HR 6318), p. 2072.

United States v. Maine, 423 United States Reporter, p. 1.

Weems, Steven L. 1975. Testimony at Boston, Mass. before the Ad Hoc Select Committee on the Outer Continental Shelf on the Outer Continental Shelf Lands Act Amendments of 1975 (HR 6218), p. 2115.

3
MAGCRC:
A Classic Model for State/Federal Communication and Cooperation

Edward Wilson

INTRODUCTION

In Dover, Delaware, the state capital, technical representatives of coastal mid-atlantic states gathered around the conference table on November 26, 1974 to share their concerns raised by the proposed outer continental shelf (OCS) exploration and development of possible oil and gas resources offshore their states. The immediate precursor to the states conference was the proposal by the federal authorities "to accelerate Outer Continental Shelf oil and gas leasing in the years 1975 through 1978. . . ." primarily by increasing to 10 million acres the 1975 area made available for OCS exploration and development. (With similar increases for subsequent years) (Final Environmental Impact Statement, p. 1).

However, antedating that action was an even more irritating concern: the long enduring litigation *United States v. Maine* to determine whether the states or the federal government had jurisdiction in these offshore areas and the right to the resources that might underlie the seabed.

Due to developing insecurity of foreign oil supplies, as a result of nationalization and expropriation, and also due to continued rising prices for petroleum resources from increased use of petroleum fuels throughout the western world, federal oil leasing activities, which had grown up in the offshore areas of the Gulf of Mexico following World War II, were to be enlarged by the sale of leases for exploration offshore the Atlantic coast. This brought an immediate response from the coastal states, many of whom claimed that the offshore petroleum resources belonged to the states rather than to the federal government and that if they were to be sold they should be sold by the states. These claims were based on original grants prior to the Revolutionary War; many grants of land also included mineral rights underneath the adjoining sea bottoms. (Coal has been mined in England from beneath the seas as early as the seventeenth century according to some accounts.) Accordingly, litigation was initiated by the state of Maine, with a number of states joining, to determine who had jurisdiction, the state government or federal government, and what the limits for development were. This case, known as *United States* v. *Maine* worked its way through the courts, finally arriving at the U.S. Supreme Court for the decision.

Although the decision had not been reached by the U.S. Supreme Court at the time of the acceleration of exploration, the decision by the Supreme Court appeared to be imminent and the federal government had decided to go ahead with some of the preliminaries to sale of leases, which took several years, on a presumption that the federal government would win. This was regarded as intransigent by the states and added to the adversary approach to the federal-state relationships.

The case was a short time later judged in favor of the federal government, but this decision increased the states' wariness rather than relieving the situation. This uneasiness was heightened by federal disclaimers that even massive discoveries and development would cause only minor onshore impacts; local officials who visited extensive petroleum-based communities in the Gulf states found these claims incredible and insisted on proof, which was not forthcoming.

The mid-Atlantic province was a frontier area where no exploration for oil and gas deposits had yet been undertaken. Geological information as to the sediments to be explored and the kerogen availability were meager to nonexistent. Among oil industry people there was considerable interest in promoting the exploration and developing the necessary information that would lead to eventual full knowledge of the availability of petroleum reserves. Some seismic and sonar-scan information had been developed, largely by proprietary interests and shared with the U.S. Geological Survey (USGS), the federal partner.

In the Gulf states oil exploration and development had been underway for several decades. Procedures were well developed and well understood by the parties involved. To a large extent this was because oil had first been discovered onshore in the Gulf coastal states and exploration had moved gradually, first into the shallow state waters and then into deeper waters, including eventually the federal waters beyond the three-mile territorial limits.

But in the mid-Atlantic states, where conflicts over the use of coastal areas were already being felt, there was concern among the states that additional conflicts would be engendered by the petroleum exploration and development. The tourism and recreation industry with vast, heavily-financed interest in the seashore was concerned about oil spills and other impacts that might have a deleterious effect on their business. The fear of the unknown (drilling and production platforms) raised concern that the vistas would be damaged. The boom and bust nature of discovery, development, and decline was a matter of concern for the local communities and the planning agencies. The possible cost of necessary state-supporting services, which might be required before any revenues were developed from the increased economic activity of the oil related projects, added additional concerns.

The federal government, particularly the Bureau of Land Management (BLM) that was responsible for the sale of OCS leases, apparently had little recognition either of the concern of the coastal states or of any possible validity in their concerns. The well developed procedures understood by those involved in the Gulf states led them to believe that impacts would be minimal. So the different perceptions of the forthcoming accelerated development of 10 million acres had elements for a somewhat adversary relationship between the federal government and the state governments.

At the Dover meeting held in 1974 various technical disciplines relating to different aspects of the forthcoming development were represented. Academic interests were involved. Some state geologists were present. Coastal zone management people were there and a number of environmental agencies were present. The meeting was held under the auspices of the Delaware State Planning Office and was identified as a "Coastal Zone/OCS Meeting" (Official "Attendance List," 1974). As the discussion of the interface between OCS exploration and coastal-zone planning and management progressed, it became evident to a number of members that at least one of the major issues was the level of support needed by the states and localities to meet the impacts of development, and the availability of resources (funds) that would be needed. Many other issues involved policy decisions, either by the federal government in their conduct of the lease sales, regulations for both the development of the

resources and the associated activities relating to the adjacent states, or those decisions related to states' basic
policies toward quick and substantial development of possible offshore resources. Conflicts based upon environmental
concerns, particularly related to damage to the flourishing
fishing industry and the estuarine resources, opposed conflicts raised by the desire of commercial developers for the
increased economic benefits of the petroleum-supporting industries that would be based in the adjacent states. General concerns with both oil spills and constraints to surface
traffic, particularly bulk carriers, were expressed. In
addition some concern was voiced regarding the basic procedures for determining what financial returns should be obtained from these resources. Was the federal government
getting adequate prices for their leases? Should leasing
and/or production revenues be shared with the adjoining
states? Would the increased petroleum traffic either increase the oil spill probability to an unacceptable level or
require pipelines that would disrupt estuarine areas as they
made landfalls? These and other questions appeared to be
beyond the authority of the planners as they contemplated
this relatively unknown activity and their responsibility to
plan appropriately for an apparent 20 to 30 year cycle. In
view of the increasing policy orientation of the issues, the
consensus was that action needed to be taken to develop
policy-oriented procedures to provide a basis for the necessary activities of the group.

After the meeting as the individual state representatives
went back to their superiors, generally at the governor or
cabinet level, and discussed the coastal/OCS issue, the general agreement was that some joint, policy-issue-oriented
action was required.

The governors acted. Hosted by Governor Byrne of New
Jersey, a meeting was set for January 7, 1975 at Princeton,
N.J. to develop strategy for the Atlantic coastal states.
Plans were also made for follow-up implementation meetings
to carry out on a regional basis the decisions of the strategy meeting. These organizational meetings, the issues developed, and the pattern of implementation furnished the
beginnings of the Middle Atlantic Governors Coastal Resources Council (MAGCRC, pronounced "Măgrăc").

THE MID-ATLANTIC GOVERNORS ORGANIZE MAGCRC

The organization and formalization of a working agreement
among a number of state governors to meet a perceived crisis

situation for which almost no previous staff work had been done took a major effort to keep from getting bogged down. The different states had worked with each other on regional projects with development usually being done at the state counterpart level, but the grouping of the governors themselves was unusual to say the least. At one strategy meeting the states of Maine, New Hampshire, Massachusetts, Connecticut, Delaware, New York, New Jersey, Maryland, Virginia, North Carolina, Pennsylvania, Rhode Island, South Carolina, Georgia, and Florida were represented by governors or lieutenant governors or their representatives, including a number of cabinet secretaries and elected state officials. In addition to the governors and their representatives, the U.S. Department of the Interior was represented by the secretary and his assistant secretary and the U.S. Congress was represented by Senator Hollings of South Carolina. Following this meeting one excellent summary and strategy analysis was contributed by the then-governor of South Carolina, who is now Secretary of Energy, the Honorable James B. Edwards.

The meetings were held in Princeton, New Jersey, Baltimore, Maryland, and Cherry Hill, New Jersey, leading to the adoption on April 10, 1975 of a resolution to the President and the Congress of the United States:

RESOLUTION
MID-ATLANTIC GOVERNORS' CONFERENCE
Cherry Hill Inn
Cherry Hill, New Jersey
April 10, 1975

WHEREAS, the Federal Government has announced its intent to lease lands on the Outer Continental Shelf for the exploration and production of oil and gas off our shores; and

WHEREAS, the Department of the Interior, acting as the administrator of the Outer Continental Shelf, has prepared a draft Environmental Impact Statement pursuant to the National Environmental Policy Act; and

WHEREAS, the impact statement prepared was deficient in its treatment not only of the environmental hazards attendant upon oil and gas exploration and production but also in its treatment of the onshore impacts, both economic and environmental, resulting from exploration and production; and

WHEREAS, it is clear from the experience of other states adjacent to off shore operations and from the experience of countries bordering North Sea operations, that such activities can have a severe and detrimental impact on the economy, environment, and social structures of affected areas; and

WHEREAS, in order to play a full and effective role in the decisions which confront them regarding the potential development of Outer Continental Shelf resources, the states and the public must be given the information necessary to make those decisions;

NOW THEREFORE, BE IT RESOLVED that the Mid-Atlantic Governor's Conference memorializes and petitions the President and the Congress of the United States to require of the Department of the Interior or such other agency as may be responsible for the preparation of an Environmental Impact Statement pursuant to Section 102 (c) of the National Environmental Policy Act (USC ___) concerning the Outer Continental Shelf leasing programs:

I. The preparation, with the full and active involvement of the States, of a detailed and accurate analysis of the onshore impact of the program.

II. The inclusion in that analysis of the short and long range economic impact of the program with particular attention to the following:

i. A 10 year plan of development of onshore and related facilities such as staging areas, rig assembly areas, pipelines, refineries, and associated industrial development.

ii. Baseline studies of the regional and local economies to be affected by off shore operations, with projections for amount and type of economic growth and development, with and without OCS development.

iii. The requirements which will be placed on state and local governments to provide support for growth induced by OCS activities, including roads, schools, hospitals, police and fire services, and other governmental infrastructure or service requirements, and the gross cost of those requirements.

iv. An assessment of the revenues which state and local government would receive from economic activity directly or indirectly caused by OCS activities, and a comparison of those revenues with the costs which would have to be borne by state and local government to support those operations.

v. An assessment of the short term economic displacements, if any, which would occur with the introduction into a state, region, or locality, of OCS-related industrial development.

vi. An assessment of the employment potential of OCS activities broken down as to both primary and secondary employment, and showing the effect on population growth in the areas to be affected.

vii. The long-term effects of conversion of areas to OCS support facilities, including an evaluation of the effects of the certain and eventual decline in OCS oil and gas production on the economic structure of regions affected.

viii. Such other areas for analysis as may be developed in consultation with the states.

III. The requirements that the evaluation of economic impact be submitted to the Governors of affected states for their review, and that it be the subject of public hearings throughout the region.

IV. That no significant actions to lease Atlantic Outer Continental Shelf lands be undertaken until such a statement is completed, reviewed by the States and the public, and found by the Governors to be a sufficient basis for decisions on Outer Continental Shelf Issues.

V. That, after tracts have been selected for leasing, that detailed and specific statements be prepared for the leasing of specific tracts, in addition to and not in lieu of, the above described programmatic statement of economic impact.

As may be seen, the major concerns of the governors were for the developing energy activities, the impact the OCS exploration and development could have on onshore infrastructures, and the state's perception that the federal government was intending to proceed with OCS development with considerably less cooperation and participation of the adjacent states than was desirable.

Thus, in addition to the legitimate state concerns outlined above, there was discussion and some feeling regarding a revenue-sharing arrangement for the adjacent states to share in the income from possible OCS hydrocarbon resources. This concern was matched by the recognition that if major resources were discovered there would be substantial economic requirements on the adjacent states to provide infrastructure and to furnish supporting services needed for the massive, sometimes boom and bust development that could possibly follow such discoveries. The budget inclusion of $2 billion of federal revenue expectations from one year of OCS leasing was regarded as one measure of the magnitude of the possible development, supporting the state's contention that adequate arrangements should be made for their sharing and participation in the OCS activities.

As a result of these conferences, the Middle Atlantic Governors' Coastal Resources Council, as it was finally designated, was formed on a quasi-regional basis, consisting of the Governors of the five mid-Atlantic states, New York, New Jersey, Delaware, Maryland, and Virginia. It is important to note that the organization papers of MAGCRC formed an agreement among the governors, not their representatives, to work together to achieve the necessary participation of the states in future resource development. This unique organization worked as an integrated advocate to take positions on related legislation and establish states' policies on federal activities, not only the activities of the U.S. Department of the Interior but any other agency or part of the government. MAGCRC worked to develop and support regional positions as necessary.

The staff work of MAGCRC was handled by two working groups, staffed by the respective states. Technical matters were handled by a technical subcommittee and policy matters were handled by policy representatives of the individual governors, generally working either in the governor's office or in the office of one of the cabinet secretaries. This

arrangement provided comprehensive consideration of technical issues and constraints and high-echelon coordination of
the policy aspects of individual issues.

MAGCRC IN ACTION

On March 17, 1975 the mid-Atlantic Governors were notified
of the unfavorable decision of the U.S. Supreme Court awarding OCS jurisdiction to the federal government. In a press
release Governor Tribbitt of Delaware stated, "My first immediate reaction was to call another meeting of the Mid-
Atlantic Governors' Resources Advisory Council . . ." The
news release went on to identify that the major effort of
the states would be to work with Congress to try to assure
that the states would participate in the decisions being
made in offshore development.

Thus the new organization of concerned states had its
first major challenge very shortly after its organization.
Working together, the mid-Atlantic states were able to develop joint positions on a number of issues and present
those concerns to the appropriate federal agency.

The federal agencies that had been involved with the
state people at the time of the organization of MAGCRC were
more inclined to accept the need for procedural changes, or
other responses when a problem area was presented not only
by several states acting together but as a joint request
from the governors of those states.

The question of the role of MAGCRC in the OCS development
was raised early in its activities. Under the cooperative
agreement with the Federal Energy Administration one of the
deliverables for the study contract was a jointly agreed
upon definition of the role of MAGCRC. After considerable
discussion the role of MAGCRC was defined.

> The Mid-Atlantic Governors' Coastal Council is a
> policy development group of Governors of the States of
> New York, New Jersey, Delaware, Maryland, and Vir-
> ginia, acting on the advice of their direct repre-
> sentatives to develop and present, as appropriate, the
> joint guidance of the five mid-Atlantic States on
> matters of policy affecting coastal resources. Major
> concerns include outer continental shelf activity by
> the Federal Government or by lessees of the Federal
> Government, which impact on State interest or on State
> activities.

The purpose as envisioned by the middle Atlantic Governors, who make up the Council, was to:

- Continually examine the position of the party States with regard to OCS management;

- Maintain consulting committees of qualified experts on scientific, technical, and administrative matters;

- Conduct such studies and prepare reports as necessary to address and identify problems of regional interest;

- Coordinate matters of interstate concern with appropriate Federal agencies;

- Coordinate the dissemination of certain public information.

In the ensuing years MAGCRC worked closely with the Department of the Interior, particularly the Bureau of Land Management but also the USGS, in the prelease activities. It participated in the tract selection procedures and contributed to both the geological and the ecological reviews involved in the leasing procedures. This relationship has continued through the years although in more recent years it has been replaced to some degree by the Regional Technical Working Group that will be discussed below.

MAGCRC/Federal Energy Administration
Cooperative Agreement

The participation of the Department of the Interior and of the Federal Energy Administration (FEA) (the percursor to the present Department of Energy) augured a good start for the new born MAGCRC. However, the Department of the Interior firmly advised that they were unable to provide any funding support for the states' new organization. The Federal Energy Administration on the other hand, as an adjunct to their efforts to evolve a national energy plan evinced a keen interest in assisting the new organization. In that context the Federal Energy Administration arranged a cooperative agreement to provide $115 million for the study of the problems of developing the outer continental shelf energy resources.

Since it was important for federal contract procedures for the FEA to deal with an established entity, the state of Delaware, under Governor Sherman W. Tribbitt, MAGCRC's acting chairman, acted as the contracting agent for MAGCRC and carried out all of the administrative details of the cooperative agreement. All of the states participated in the activities equally, sharing in the work and costs of the agreement but working as subcontractors to the state of Delaware each with a separate contract with Delaware for their services. This arrangement worked out with no problems and all of the agreed upon deliverables were produced.

Under the cooperative agreement MAGCRC furnished analysis of six major studies, all related to socioeconomic and other impacts of the proposed OCS development. It also developed individual studies covering all of the legislative and regulatory procedures in the individual states that would interface with the OCS activities. The basic thrust of the project was to identify management decision points and provide a preliminary assessment of the information that would be needed for energy and OCS planning. Also to be considered were plans for a combined regional system to manage onshore impacts and timing of OCS deepwater ports and related activity.

The existing OCS studies that were to be analyzed were as follows:

1. *OCS Oil and Gas--an Environmental Assessment,* prepared by Resource Planning Associates, Inc. (RPA) for the Council on Environmental Quality (CEQ), and published in April, 1974. RPA, the major contractor for this study, investigated onshore impacts in the Mid-Atlantic region and three other frontier regions (two in the Atlantic and one in the Gulf of Alaska). Other contractors examined relevant OCS technology, offshore oil spills, and the ecology of the regions.

2. *A Socioeconomic and Environmental Inventory of the North Atlantic Region,* presented to the Bureau of Land Management by The Research Institute of the Gulf of Maine (TRIGOM) in November 1974. The inventory covers the coastal states from Sandy Hook, New Jersey, to Bay of Fundy, Maine.

3. *A Study of the Socioeconomic Factors Relating to the Outer Continental Shelf of the Mid-Atlantic Coast,* prepared by the College of Marine Studies (CMS), University of Delaware, for the Bureau of Land Management, and distributed in July, 1975.

4. *Mid-Atlantic Regional Study--An Assessment of the Onshore Effects of Offshore Oil and Gas Development,* conducted by Woodward-Clyde Consultants for the American Petroleum Institute (API), and published in October 1975.

5. *Draft Environmental Statement for Proposed Outer Continental Shelf Mid-Atlantic Oil and Gas Lease Sale (No. 40),* released by the Bureau of Land Management in December 1975; and Technical Paper Number 1, *Economic Study of Possible Impacts of a Potential Baltimore Canyon Sale,* New York OCS Office, December 1975.

6. *A Study of the New Use Demands on the Coastal Zone and Offshore Areas of New Jersey and Delaware,* prepared by Braddock, Dunn and McDonald. This study is not yet complete; draft materials are being reviewed and revised by the Congressional Office of Technology Assessment (OTA). The final report, which will also address impacts of other OCS technologies (e.g., deepwater ports and floating nuclear plants), will not be available until spring 1976.

A "Request for Proposal" was prepared jointly by the MAGCRC technical representatives, bids were received from a number of eminent consulting organizations, and, after individual and joint analysis by the MAGCRC study panel, Resource Planning Associates, Cambridge, Massachusetts was selected as the contractor, based upon its presentation and also its experience in preparing one of the studies related to the subject at hand.

In due course the *Identification and Analysis of Mid-Atlantic Onshore OCS Impact* was produced and accepted by both the MAGCRC technical people and the Federal Energy Administration contract people.

Following this the individual state analysis covering in great detail the many laws and regulations affecting OCS-related activities were compiled and submitted.

During the conduct of these studies, the analysis of the findings, and working with the areas of concern, the mid-Atlantic states became familiar with the potential problems associated with OCS development, becoming more comfortable with some of the areas that had previously caused concern but substituting an equal number of new concerns.

The support of the Federal Energy Administration was critical to the success of the new organization as it provided a common activity plus a comprehensive project that brought the states technical people and many of the counterpart federal people together on working terms.

During this period the Supreme Court came to a decision on the litigation *United States* v *Maine,* awarding jurisdiction of the outer continental shelf lands to the federal government. This is discussed in the following section.

MAGCRC and the OCS Advisory Board

Meanwhile, the Department of the Interior was moving to develop a nationwide forum of state representatives that would act in a fashion somewhat similar to that of MAGCRC. This would be an advisory board to the Secretary of the Interior, formally the OCS Advisory Board, which would have the single function of offering recommendations to the Secretary of the Interior. This board of course was constrained much more than the MAGCRC group that was also broadly involved in legislative activities for both the OCS Lands Act Amendments of 1978 and the Coastal Zone Management CZM Amendments of 1976.

Although the OCS Advisory Board to a degree supplanted MAGCRC, due to its advisory constraint, and also due to the fact that recommendations by the Advisory Board did not always completely meet the full needs of the concerned states, MAGCRC was continued as an organization by the mid-Atlantic Governors.

On issues that were particularly critical or of high priority to the mid-Atlantic states and particularly on legislative issues MAGCRC continued to hold meetings, to develop positions as needed, and to work through state channels to mitigate or otherwise modify actions that were undesirable from the states' viewpoint.

In this activity MAGCRC could work with the BLM Regional Office, with the upper echelons of the Department of the Interior in Washington, with the Department of Energy, or through their legislators in the Congress in preparing and influencing legislation. A number of MAGCRC meetings were attended by congressional staff members during discussions of pertinent legislation on both OCS and coastal-zone management. MAGCRC maintained a relatively low profile; more could be accomplished working directly with the decision makers than through channels involving a number of layers of government.

MAGCRC LEGISLATIVE ACTIVITIES

In 1975 Congress considered the passage of major legislation affecting OCS activities via two different approaches.

Although several bills were proposed, S-521 to amend the OCS Lands Act and S-586 to amend the Coastal Zone Management Act pursued parallel courses and were the targets of much of this activity.

A major concern of the coastal states was the direct cost to them of either providing the necessary onshore facilities for the support of offshore exploration and development or the other governmental activites, such as environmental management, scientific research, regional planning activities at both the state and local level, and protection of marine resources for both environmental and fisheries reasons. To this end the states were seeking some form of revenue sharing through which the federal government would provide necessary funding for the state activities generated specifically by OCS activity.

Although the Coastal Zone Management Act had been passed early in the decade, progress had not yet brought about the development of effective Coastal Zone Management programs in many of the coastal states. Program planning prior to federal approval and implementation of programs was the rule rather than the exception. Thus the Coastal Zone Management planning activities were relatively little help to the OCS planning activities at that time.

The Coastal Zone Management planning funds were handled through the Department of Commerce while the OCS activities were handled through the Department of the Interior, so there was not a close coordination of these two types of activities to provide OCS support; in fact there were many areas in which conflict between the two federal departments was quite possible.

Within the individual states OCS activities were handled by various institutional arrangements. Some states included OCS activities in their Coastal Zone Management planning. Some states included OCS activities as environmental concerns. And some states handled OCS activities by somewhat autonomous agencies, such as developmental or commercial activities.

So for a number of reasons both bills in Congress included sections that would provide funding specifically for OCS activities, primarily for planning purposes but with some latitude for environmental mitigation.

The Coastal Zone Management amendments were passed by Congress, but, due to substantial and effective industry opposition among other factors, the OCS Lands Act Amendments were returned to committee. Thus the OCS funding mechanism in the CZM program became available in the form of the Coastal Energy Impact Program (CEIP).

In practice this was effective in some states where Coastal Zone Management programs were becoming effective and were working with OCS programs. However in other states that had different organizational patterns or in states where Coastal Zone Management programs were not progressing satisfactorily the Coastal Energy Impact Program funds did not meet the needs of the OCS activities programs.

Meanwhile, Sale 40, the Bureau of Land Management's first sale of leases in the mid-Atlantic frontier area, was held towards the end of the year. However, with the failure of passage of the OCS Lands Act Amendments, which were support- ed by the adjacent coastal states, there was a substantial dissatisfaction with the level of protection available for this sale. The sale was challenged in court and was effec- tively halted. The MAGCRC members participated in discus- sions and activities related to blockage of this sale.

With support from the MAGCRC members and other coastal states the OCS Lands Act Amendments legislation was reintro- duced in Congress the following year. The legislation was passed and the court suit blocking Sale 40 was resolved so that exploration in the mid-Atlantic area could proceed. While it would be impossible to demonstrate any single con- nection between the problems with Sale 40 and the passage of the legislation, it becomes quite clear in reviewing the various actions by parties involved that the participation and cooperation of the coastal states can be a major factor in expediting OCS activities by the federal government.

MAGCRC TODAY

After passage of the OCS Lands Act Amendments all parties concerned worked fairly well together to set up procedures to implement the mandates of the legislation. Due to the involvement of the coastal states and the cooperation of the federal agencies there was fairly good understanding of what was needed and the operating agencies moved to do what was necessary. Most of the provisions of the OCS Lands Act Amendments worked reasonably well thus minimizing the prob- lems between the coastal states and the Department of the Interior.

One area, in which the Department of the Interior did not have jurisdiction, has to this date not been adequately re- solved to meet the intent of the law. An OCS Administrative Fund was set up to provide funding support for strictly OCS activities, particularly in those states where they were not part of the Coastal Zone Management program. This funding was provided since in some states the division of duties between OCS and Coastal Zone Management programs was incompatible with using coastal zone money (Coastal Energy Impact Fund) for the necessary supporting activites related to OCS. However, for reasons too arcane to discuss here, the operation of the OCS administrative fund, which was specified as a separate fund from Coastal Zone Management to meet the needs of those states for whom the funding was provided, was assigned to the Department of Commerce. Their Coastal Zone Management Office, which was at the time endeavoring to use all means within its power to entice the coastal states into Coastal Zone Management programs, through its legal service read the provisions in such a way that they were able to require the participation of a given state in the Coastal Zone Management program before granting them eligibility for the OCS funding. This controversy has not been settled to date to the satisfaction of the states involved. Other than the continuing effort for resolution of that problem, procedures had been reasonably well implemented to form a good and effective working relationship between the federal government and the coastal states with regard to OCS activities.

MAGCRC has been active in the recent years in the overview of the implementation of the OCS Lands Act Amendment. As controversy has been at a minimum there has not been any overt action of the organization needed.

With the new Reagan administration in Washington substantial changes to the OCS program have been proposed. The states have not been consulted in the preparation of these changes.

REFERENCES

Final Environmental Impact Statement (E.I.S.), "Proposed Increase in Oil & Gas Leasing on the OCS," vol. 1 of 3, p. 1.
Official "Attendance List", meeting of November 26, 1974.

PART III

THE
FEDERAL
PERSPECTIVE

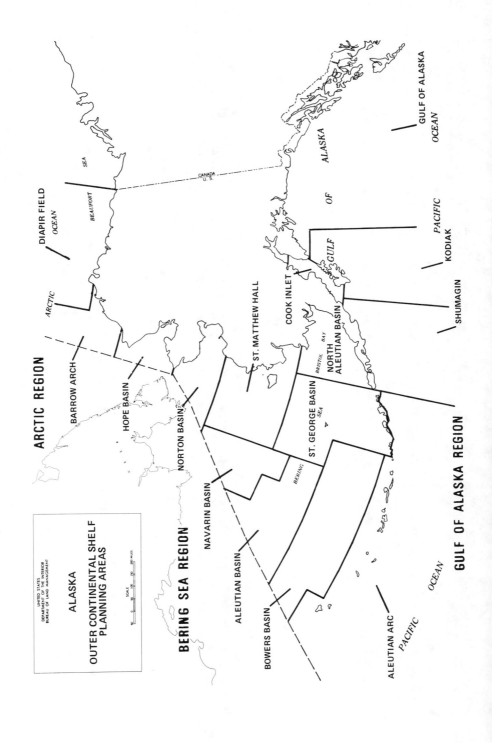

UNITED STATES
DEPARTMENT OF THE INTERIOR
BUREAU OF LAND MANAGEMENT

ALASKA

**OUTER CONTINENTAL SHELF
PLANNING AREAS**

SCALE

ARCTIC REGION

DIAPIR FIELD

BARROW ARCH

HOPE BASIN

NORTON BASIN

BERING SEA REGION

NAVARIN BASIN

ALEUTIAN BASIN

BOWERS BASIN

ALEUTIAN ARC

ST. MATTHEW HALL

ST. GEORGE BASIN

NORTH
ALEUTIAN BASIN

COOK INLET

KODIAK

SHUMAGIN

GULF OF ALASKA
OCEAN

GULF OF ALASKA REGION

ARCTIC
OCEAN

BEAUFORT SEA

CANADA
U. S.

ALASKA

GULF OF

PACIFIC

PACIFIC
OCEAN

BRISTOL BAY

BERING SEA

4

The Challenge of the Alaska OCS

Esther C. Wunnicke

THE CHALLENGE OF THE ALASKA OCS

The challenges presented in a program to develop the energy resources of the offshore areas of Alaska are great. The magnitude, diversity, and frontier nature of Alaska's outer continental shelf (OCS) and the potential resources it represents combine to make the OCS program one of the most important and exciting works of this decade in support of President Reagan's program to make the nation far less dependent upon imported oil.

The U.S. Geological Survery has estimated that there are 7 to 32 billion barrels of oil and 30 to 97 trillion cubic feet of gas undiscovered and recoverable off Alaska's shore. "Undiscovered" is the key word. There have to date been no discoveries on the outer continental shelf offshore Alaska. Offshore developments in the upper Cook Inlet and the Beaufort Sea have been on state-owned submerged lands.

It is common in the oil and gas industry to call a wildcat area a "frontier area." Offshore Alaska is a frontier area in more than that sense. Other portions of the nation's outer continental shelf are frontier areas in that no oil and gas have been discovered and no production developed. But onshore on the Atlantic coast, for example, there are well-developed support and transportation systems to handle the oil or gas once they are found. Not so in Alaska. Most of the over 45,000 miles of shoreline are undeveloped in character. Once one leaves Anchorage and the Kenai Peninsula, most of the coastal communities are small, rural, and predominantly inhabited by Alaska Eskimos, Indians, and

Aleuts. There are no roads, pipelines, or established support facilities. In addition, there are few regional governments in much of coastal Alaska and there is grave concern on the part of the residents of the small coastal communities that they may be overwhelmed by imported residents associated with the search for and production of oil and gas off Alaska's coast. A sensitivity to that concern has already been evinced by the manner in which the exploratory work in the first Gulf of Alaska sale was conducted. Workmen were flown to the exploratory wells from as far away as Seattle; their only stop near the affected community was at the airport. When weather prevented transfer from the airport to the drilling rig, crews were bunked and fed in an enclave established by the oil-well operator and were forbidden by their employers to enter the village. This helped eliminate a sudden disturbance of the traditional social structure and culture of the community.

Working without established facilities will present both a challenge to and an opportunity for successful bidders. Construction of those facilities will be a part of the cost of operation for operators wherever they are located. Often the opportunity will exist to respect local requests in choosing their location. The already constructed trans-Alaska pipeline now pumping oil to the ice-free port at Valdez and the promise of construction of the Alaska-Canada gas pipeline, added to the high oil and gas potential of the Diapir Field in the Beaufort Sea, make lease offerings in that portion of the offshore Arctic especially attractive. It is the only area where transportation facilities are in place and support facilities have been established by operators of the onshore Prudhoe and Kuparuk oil fields already existing.

The scale of Alaska is difficult to fully grasp even for those who live and work there. The Alaskan outer continental shelf nearly matches in area the land mass of Alaska. This measurement is made from the three-mile territorial sea to the 200 meter water depth and excludes the offshore areas of southeast Alaska. The 51 million acres in the Diapir Field planning unit are just larger than the State of Nebraska, and the Navarin and North Aleutian Shelf, each at 33 million acres, are each half as large as the state of Colorado. In the St. George unit, over twice as big as the Navarin, there has been one continental offshore stratigraphic test (COST) well drilled to date. One can imagine the difficulty of projecting potential oil and gas resources in an area the size of Colorado based on the downhole information of a single test well.

With estimates high but actual knowledge of oil and gas resources slight, successful ·bidders at Alaska OCS lease sales are confronted with a diversity of drilling and production conditions. No two areas are identical. The winds, waves, and storm surges of the Gulf of Alaska are matched by the ice conditions in the Beaufort and Chukchi Seas as tests of industry's technology. Short open-water exploratory drilling seasons in the Bering Sea, for example, and the winter temperatures and darkness test the mettle of all dealing with Arctic Alaska. No other offshore region has the ice hazards found in the Beaufort and Chukchi Seas off Alaska. The subsistence lifestyle of coastal residents-- depending in most instances on the fish and marine mammals of the ocean--is unique to rural Alaska. The great populations of birds, fish, and marine mammals and large numbers of endangered whale species migrating through the Alaska outer continental shelf (some are permanent residents) intensify the issue of proposed oil and gas development.

History

Like most controversial issues, Alaska outer continental shelf leasing has often been challenged in the courts. The 1953 Outer Continental Shelf Lands Act (The Outer Continental Shelf Act of 1953, 43 USC 1331-1356), establishing federal jurisdiction over the submerged lands of the outer continental shelf, was passed by Congress six years before Alaska was admitted as the forty-ninth state. Twenty years after passage of that act, the Bureau of Land Management put an environmental assessment team in Alaska. The first Alaska OCS sale was held for the Gulf of Alaska in April 1976, but not before the State of Alaska and the village of Yakutat sought a preliminary injunction to delay the sale (Alaska v. Andrus, 1978). The injunction was denied and the sale was held with later declaratory relief from the Court of Appeals in 1978, which directed consideration of alternative operating orders. The issues raised by this suit were discussed and were a part of the impetus for the passage of the OCS Lands Act Amendments (Outer Continental Shelf Lands Act Amendments, 1978, 43 USC 1801-1866) in 1978. In addition to establishing policies to expedite exploration and development of the outer continental shelf, the Act directed the Secretary of the Interior to do so in a manner that would:

. . . preserve, protect, and develop oil and natural
gas resources in the Outer Continental Shelf in a manner which is consistent with the need (A) to make such
resources available to meet the Nation's energy needs
as rapidly as possible, (B) to balance orderly energy
resource development with protection of the human,
marine, and coastal environments, (C) to insure the
public a fair and equitable return on the resources of
the Outer Continental Shelf, and (D) to preserve and
maintain free enterprise competition; . . . [Outer
Continental Shelf Lands Act Amendments, 1978, 43 USC
1802]

Thirty-nine bidders vied for 81 of the 189 tracts offered
in what held promise of a giant oil or gas field between
Kayak Island and Icy Bay. Eleven wells were drilled in the
five-year lease period. No discoveries were made and neither the promised bonanza of the industry nor the feared
evironmental and social disruption of the state were realized. Four years later, in October 1980, when a second Gulf
of Alaska sale (Sale 55) was held closer to Yakutat, voices
were muted. Bidding interest was slight, with only 37 of
the 210 tracts offered receiving bids, and opposition to the
sale was significantly reduced. There was no litigation.
Merely determining federal and state boundaries and jurisdiction in the Cook Inlet occupied the courts from 1967
to 1975, when the U.S. Supreme Court ruled that Cook Inlet
was not an historic bay as the State of Alaska had contend-
ed. The closure line of the boundary, the court decided,
followed the 24-mile international closure rule for bays.
This placed the line at Kalgin Island with the territorial
sea to be measured from that line (U.S. v. Alaska, 1972).
With jurisdiction determined, an OCS sale was scheduled for
February 1977 for Cook Inlet south of Kalgin Island. In-
junctive relief from the proposed sale was sought in January
by the village of English Bay and others. The sale was voluntarily postponed, but only until September. By March, the
plaintiffs had resumed their suit. An amicable settlement
of the suit was later reached and the case was dismissed
with prejudice (English Bay Village Corp. v. Secretary of
the Interior, 1977). The stipulations agreed upon in settlement of the English Bay suit have since been considered
in succeeding Alaska sales. When the Cook Inlet lease offering was held October 27, 1977, 31 bidders bid on 91 of
the 135 offered tracts, and five-year leases were issued on
87 tracts. Ten wells were drilled. All ten were subse-
quently plugged and abandoned. No further exploratory wells

are anticipated before the December 1, 1982 expiration date.

Opposition to offshore drilling in the Cook Inlet area was not limited to the federal outer continental shelf. In 1973, the State of Alaska issued oil and gas leases in Kachemak Bay and fishermen and environmental organizations joined to challenge the legality of the sale (*Moore v. Alaska, 1976*). In response to this challenge, a marine sanctuary was established by the Alaska legislature in Kachemak Bay, and repurchase of all major leases was negotiated by the state at a cost of $28 million.

On September 29, 1981, those tracts not leased in the original Cook Inlet lease sale, as well as other tracts farther south in Cook Inlet and extending well into Shelikof Straits, were offered in OCS Sale 60. Many groups expressed opposition to the offering of tracts, especially those in Shelikof Straits. Commercial and subsistence fishermen in particular opposed the sale. Although the state of Alaska did not oppose the sale, it did propose major modifications of the area to be offered and of the lease terms.

In response to state concerns, lease terms contained a wellhead and pipeline design stipulation to further reduce chances of fouling fishing gear on underwater facilities. The state is also to be involved in implementation of a stipulation designed to protect biological resources.

Like the Gulf of Alaska, prior drilling experience in lower Cook Inlet dampened bidding enthusiasm. Only two companies participated in the sale. Only 13 of the 153 tracts offered received bids and were leased. It was one of the smallest lease sales in the history of the entire OCS leasing program from the standpoint of bonus monies received and bidders participating.

The joint Beaufort Sea sale held in December 1979 was unique in OCS history. It was the first lease sale in the American Arctic and the first joint sale with a coastal state. A boundary dispute between the United States and the State of Alaska mandated the joint nature of the lease sale (*U.S. v. Alaska, 1979*). Among the boundary issues to be settled by the Supreme Court upon report of a Special Master is the determination of whether the shoreline is to be measured by the sinuosities of the coast or from point-to-point on a so-called straight baseline concept. Rather than postpone development of an area of high potential and one, as noted previously, near an already established oil pipeline, the two governments agreed to a joint sale and to judicial determination of the ownership of the disputed tracts. Four of the 23 undisputed federal tracts were leased in the sale, but federal leases were issued on 20 disputed tracts. The major part of the lease sale--58 tracts--involved nearshore tracts in undisputed state ownership. Four tracts leased from the state are in the disputed Dinkum Sands area. Dinkum Sands is claimed by the state as an island within its own three-mile territorial sea and is critical as a "connecting point" in the straight baseline concept. It is a formation viewed by the federal government as submerged lands and is thereby a part of the outer continental shelf.

Injunctions were sought by the North Slope Borough, the National Wildlife Federation, and the village of Kaktovik against conducting the sales (*North Slope Borough v. Andrus, 1980a*). Although the preliminary injunction was denied and the sale allowed to go forward, the Federal District Court enjoined acceptance of bids, issuance of leases, and all activity on federally-managed tracts pending a new environmental impact statement. The District Court was reversed on appeal and a petition for rehearing was denied (*North Slope Borough v. Andrus, 1980b*).

Nine bidding entities participated in the sale and the total bids accepted by both governments exceeded a billion dollars. Bidding systems used in the sale included bonus bidding and "fixed net profit" share on state tracts. In recognition of severe constraints presented by weather, ice, and darkness, as well as the reduced exploratory drilling season and test structure requirements of lease terms themselves, ten-year leases were issued by both governments. Although allowed to issue leases early in 1980, the state of Alaska also faced litigation brought by the North Slope Borough and the village of Kaktovik (*North Slope Borough v. Alaska, 1980*). The Alaska Superior Court enjoined activities on state leases beyond the Barrier Islands but the injunction was stayed by the Alaska Supreme Court. The Alaska Supreme Court heard arguments in March of 1981, but by January 1982 had not ruled on the merits of the case. Among the issues raised was that coastal zone management consistency determinations had not been adequate. As one of the few large regional governments in coastal Alaska, the North Slope Borough has developed a coastal-zone plan and adopted a coastal-zone management ordinance. Other coastal areas in the unorganized Borough of Alaska are participating in statewide coastal-zone management by establishing Coastal Service Resource Areas which are to develop plans and ordinances for the coastal zone.

Following the precedent set in the North Atlantic, a Biological Task Force was required in the joint federal/state Beaufort leases. Two other stipulations--one requiring test structures and the other a so-called seasonal stipulation

limiting downhole activities to a five-month winter period--
were without precedent in earlier OCS leasing. But leasing
in the American Arctic OCS was also without precedent. The
seasonal stipulation was adopted in response to a biological
opinion given by the. National Marine Fisheries Service concerning bowhead whales. The National Marine Fisheries
Service is the agency charged with the responsibility of
protecting endangered species under the Endangered Species
Act (Endangered Species Act of 1973). The stipulation was
made a part of both federal and state leases. The federal
stipulation presently states:

> Exploratory drilling and testing and other downhole
> activities will be limited to the period November 1
> through March 31 unless the Supervisor determines that
> continued operations are necessary to prevent a loss
> of well control or to ensure human safety. This stip-
> ulation will remain in effect for two years following
> the issuance of the lease. [Department of Interi-
> or/State of Alaska, 1979, 44 FR 64762]

The state stipulation for the same purpose states:

> Exploratory drilling and testing and other downhole
> exploratory activities from surface locations outside
> the Barrier Islands will be limited to the period
> November 1 through March 31, unless the Director, Di-
> vision of Minerals and Energy Management, after con-
> sulting with the Oil and Gas Conservation Commission,
> determines that continued operations are necessary to
> prevent a loss of well control or to ensure human
> safety. This stipulation will remain in effect for
> two years following issuance of the lease. [Department
> of Interior/State of Alaska, 1979, 44 FR 64765]

In addition, the Information to Lessees from the Sale Notice
provides that:

> Federal Stipulation No. 8 has been developed to pro-
> vide interim protection for the˙ endangered whales dur-
> ing this 2-year period. At the end of this period,
> lessees will be advised as to what, if any, restric-
> tions on operations will be necessary to be consistent
> with the Endangered Species Act. This determination
> will be made by the Secretary of Interior in consulta-
> tion with NMFS. [Department of Interior/State of Alas-
> ka, 1979, 44 FR 64768]

The Information to Lessees section also specifies that the Bureau of Land Management (BLM) would, in order to comply with the Endangered Species Act, collect additional information for assessing possible impact on bowhead whales. To this end, BLM conducted extensive research during the past three years on populations, migration routes, and behavior of bowhead whales. Many of the studies were conducted in the Canadian Beaufort where offshore exploratory drilling has been taking place for several years during both summer and winter seasons. Following consultation with the National Marine Fisheries Service and based upon information developed through the BLM studies program, both the Secretary of the Interior and the governor of Alaska will review the seasonal stipulation to determine whether it should be continued, removed, or modified. Use of the bowhead for food by the Inupiat people of the Alaska Arctic and the place of the hunt and community sharing of the kill as a part of Inupiat culture combine with the status of the bowhead as an endangered species to point up the extreme sensitivity of these forthcoming decisions. Beginning in 1978 and continuing throughout the two years following the joint Beaufort sale, many studies have been funded and conducted by the Alaska OCS Office of the Bureau of Land Management to facilitate future decisions about endangered whales. The whale studies are part of a larger environmental studies program.

The promise of oil and gas in the joint-lease area and in the Diapir Field in the Beaufort Sea justifies the expense and effort to determine federal-state boundaries and ownership. It is estimated that the tidal shoreline of Alaska, including islands and inlets, totals 47,300 miles. It is probable that, as in the Beaufort Sea, other questions as to whether the state or the United States has the right to lease certain offshore areas will arise. Determination of coastal boundaries between the United States and the state of Alaska is in itself a herculean task. International boundaries with neighboring Canada in the Arctic and the Soviet Union in the Bering Sea give an added dimension to the search for oil and gas resources on the Alaska outer continental shelf.

Process

The Secretary of the Interior is directed in the Outer Continental Shelf Lands Act, as amended, to conduct a studies program to "establish information needed for assessment and management of environmental impacts on the human, marine, and coastal environments of the outer continental shelf and the coastal areas that may be affected by oil and gas development" (The Outer Continental Shelf Lands Acts of 1953, 43 USC 1346). Since 1974, research and study to that end in Alaska have accounted for more than half of the $260 million spent nationwide by the Bureau of Land Management for OCS studies. This attention to the Alaska OCS results not only from its size (over 70 percent) in relation to the total for the nation, but also from the existence in offshore and coastal Alaska of extensive populations of birds, fish, and marine mammals. For example, eight species of whales, five of them endangered, either live year-round in or migrate through Alaskan waters. Marine mammals—four species of seal, walrus, and polar bear—abound in offshore Alaska. As a case in point, the Alaska ports of Dutch Harbor and Kodiak rank first and second, respectively, in the United States in the value (gross earnings to fishermen) of fish brought to port.

The focus of the studies program is to provide information to decision makers who guide the OCS leasing program. In addition, the Alaska scientific community and other entities have benefited from research done and knowledge gained through the environmental studies program. Overall policy, management, planning, and budgeting responsibility rest with Bureau of Land Management headquarters in Washington, D.C. Regional study needs are developed in the Alaska OCS Office and combined with national needs for each year's study program. Although some Alaska studies are individually contracted, the majority of the environmental studies are designed and monitored under a basic agreement between the Bureau of Land Management and the National Oceanic and Atmospheric Administration (NOAA). Through this NOAA Outer Continental Shelf Environmental Assessment Program (OCSEAP) based in Juneau, Alaska, study results are also synthesized and put in readily useable form for those who do the impact analysis and prepare environmental statements and other documents for decisions regarding the size, timing, and conditions of proposed lease offerings.

Those responsible for the Alaska OCS program are very conscious of the totality of the Alaska environment and its human elements. The potential socioeconomic impacts of OCS oil and gas development are also being explored. This socioeconomic studies program assembles predevelopment information and monitors project development as it affects specific communities, regions, or the state as a whole. With such a small population base (the state has a population of less then a half million) and an economy based for the major part on primary extraction of renewable and nonrenewable natural resources, extreme care and sensitivity are required in the projection of the social and economic effects of OCS development. Some of those effects may be difficult to separate from the consequences of a number of other events combining in an unsettled brew in Alaska.

Dramatic distributions of public lands from the federal government to native corporate and private owners, as mandated by the Alaska Native Claims Settlement Act (Alaska Native Claims Settlement Act, 1971), are being made. Transfers to the state of Alaska of lands promised as a part of the Alaska Statehood Act (Alaska Statehood Act, 1958) only 23 years ago are also being made. Revenues from state lands have already been acquired, most notably from oil and gas development at Prudhoe Bay, fuel capital improvements, and other state programs that also have social, economic, and environmental effects. Planned and executed in advance of proposed leasing, environmental and socioeconomic studies are the foundation for analysis and leasing decisions.

The BLM's Alaska OCS and Washington offices are responsible for preparation of environmental impact statements (EIS) in advance of final sale decisions. One team in the Environmental Assessment Division devotes its time to analysis of potential impacts of leases in Arctic waters and has built up a body of expertise on the conditions and issues peculiar to areas in the northern Bering, Chukchi, and Beaufort Seas. Another team has concentrated on obtaining information, identifying issues, and assessing potential impacts of proposed offshore drilling in the southern Bering Sea and Gulf of Alaska regions.

Throughout the leasing process, extensive coordination takes place between the OCS office, other agencies, private groups, state government agencies, and local communities. Meetings to develop the scope of the EIS are held throughout Alaska as part of the effort to provide information to outlying communities as well as to gain knowledge of their concerns and suggestions for issues to be considered. Formal public hearings on draft environmental impact statements are important to assure that all potential impacts are accurately depicted. Involving people from the affected communities requires the best efforts of the Alaska OCS Office. Although constrained by weather and distance, all possible ways to permit involvement of local people are sought. When hearings were held in July on the EIS for the proposed five-year schedule, the state of Alaska's teleconference network was employed to allow the hearing panel to receive testimony from citizens in 26 locations around the state. There is always great interest in meetings or hearings involving the OCS program. For example, at hearings held in Nome, Bethel, Unalakleet, Emmonak, Savoonga, Kotlik, and Anchorage in early October, 1981, 170 persons contributed oral testimony on the draft environmental impact statement (DEIS) for proposed sale 57, Norton Basin, and 40 more provided written testimony.

In addition to cochairing the Regional Technical Working Group that advises the OCS Manager for Alaska and the director of the Bureau of Land Management, the state of Alaska is formally notified of and participates in all public steps of the OCS leasing process.

The entire process is one which attempts to base decisions on scientific fact, good analysis, and full public participation. It is the Alaska OCS office's challenge to retain those hallmarks of the process in the lease offerings proposed off Alaska. The 1980 schedule proposed ten lease sales on the Alaska outer continental shelf. A draft schedule issued in July 1981 calls for 16 Alaska lease sales in the next five years. Specifics of that schedule are not final at this time, however, because of the October 1981 court ruling against the five-year schedule adopted in June 1980 by then Secretary Cecil Andrus (*California et al. v. Andrus, 1981*). In response to that ruling, Secretary James Watt is reviewing this proposed schedule and the supporting decision documents to assure that the "balancing" determinations and additional public participation required by the court are achieved. Announcement of the final approval of a new five-year schedule will be made upon the advice of the Court of Appeals sometime in 1982.

Future

Forecasting the level of oil and gas activity on the Alaska outer continental shelf is speculative at best. Even forecasting discoveries is difficult and production after discovery may take years to achieve. To make those yet undiscovered resources available with minimum adverse impact to the lands and waters of Alaska and the people who live there is the challenge. To meet the challenge requires the best efforts of government, industry, conservationists, developers, and Alaska's people.

REFERENCES

Alaska Native Claims Settlement Act, 85 Stat. 688 (December 18, 1971): 43 U.S.C. 1601 et seq.

Alaska Statehood Act, 72 Stat. 339 (July 7, 1958): 48 U.S.C. 2.

Alaska v. *Andrus.* 1978. 580 F. 2d 465 (U.S. Ct. App., D. of Columbia).

California et al. v. *Andrus.* 1981. Civil Nos. 80-1894, 1897, 1935, and 1991, U.S. Ct. App., D. of Columbia, decided October 6.

Department of Interior/State of Alaska. 1979. Joint Federal/State Beaufort Sea Oil and Gas Lease Sale BF, Final Notice of Sale, 44 FR 64762 (November 7, 1979).

Endangered Species Act of 1973, P.L. 93-205, 87 Stat. 884 (December 28, 1973): 16 U.S.C. 1531-1543, as amended.

English Bay Village Corp. v. *Secretary of the Interior.* 1977. CA 77-0174 (U.S. District Court, D. of Columbia).

Moore v. *State of Alaska.* 1976. 553 P. 2d 8 (Alaska).

North Slope Borough v. *Alaska.* 1980. U.S. Supreme Ct. No. 5550, and consolidated cases 5558, 5560, and 5561 (Superior Ct. 2BA-79-57 Civ).

North Slope Borough v. *Andrus.* 1980a. 486 F. Supp. 332 (U.S. District Court, D. of Columbia).

North Slope Borough v. *Andrus.* 1980b. 642 F. 2d 589 (U.S. Ct. App., D. of Columbia).

The Outer Continental Shelf Lands Act of 1953, 67 Stat. 462 (August 7, 1953).

Outer Continental Shelf Lands Act Amendments. 1978. P.L. 95-372, 92 Stat. 629 (September 18, 1978).

United States v. *Alaska.* 1972. 352 F. Supp. 815 (U.S. District Court, Alaska), 497 F. 2d 1155 (U.S. Ct. App., 9th Cir. 1974); reversed and remanded 422 U.S. 184, 45 L. Ed. 2d 109, 95 U.S. Supreme Ct. 2240 (1975).

United States v. *Alaska.* 1979. U.S. Supreme Ct. No. 84, Original.

5
Environmental Issues and Offshore Oil and Gas: Mid-Atlantic Region

Frank Basile
Barbara Karlin

Each of us has a subjective affinity for what we consider our environment. We may define our environments differently, but we almost always include these perceptions in the definition: (1) our particular environment is unique and intrinsically valuable; (2) it is shared, sometimes unwillingly, with others; and (3) change is suspect, and any proposal that changes our share is intolerable.

Given those perceptions, it is small wonder that the first attempts in the mid-1970s to expand exploration for offshore oil and gas outside of the Gulf of Mexico were met with so much skepticism and resistance. To be fair, skepticism can serve a useful function--it limits the pace and breadth of any change. So the first attempts to expand oil and gas development into offshore frontiers raised, to some extent justifiably, the level of public skepticism. The infamous Santa Barbara oil spill of 1969--and a nationwide study in 1969-70 to locate deepwater ports for importing seemingly unlimited quantities of cheap foreign oil--set the stage for increased local resistance. The fact that several offshore lease sales were scheduled near the most densely populated section of the country caused people to resist more tenaciously.

At first, two negative themes pervaded the reactions to potential oil development off the Atlantic coast: the certainty of oil spills; and the injustice of proposing drilling for oil and gas without a national energy policy in place. Over time, the negative themes grew in number.

It was thought at first that any large oil spill inherently threatens to destroy tourism and permanently foul

103

bottom sediments and beaches. Such a spill would also destroy or degrade finfish and shellfish stocks by genetically altering behavior, fertility, or resistance to disease. The spread of oil would also kill larval fish, plankton, and other organisms. Finally, oil spills were thought to be a direct threat to human health because the oil becomes encapsulated, along with diseased organisms from discharged wastes, in tar balls, which wash up on beaches.

Then entire new categories of potential disasters arose. An influx of hundreds of thousands of oilfield workers and support people would cause untold sociological and socioeconomic disruption. Support bases for offshore oil and gas would occupy thousands of coastal acres. Refineries, vast petrochemical complexes, and other facilities would displace local homes and businesses. The appearance or expansion of offshore drilling in areas previously free of drilling became a revolutionary event that threatened established patterns and functions in coastal communities along the Atlantic Coast.

In addition, this country's lack of a publicity-articulated, comprehensive strategy to break the stranglehold of foreign nations that manipulate the price and availablility of oil caused a complex set of expectations to evolve. These expectations quickly clouded the picture of the need for increased domestic production. Our need to offset, as rapidly as possible, foreign control over major sources of conventional hydrocarbon fuels became a competition among advocates of different technologies to completely revise America's energy sources and to develop an entirely new energy supply system. If nothing else, the national debate over energy sources and uses made difficult choices almost impossible because it focused attention on the question of what effect offshore oil and gas activities might have.

In the climate that developed during the 1970s debate over expanding offshore oil and gas operations into new areas, the historical users of the Atlantic outer continental shelf (OCS) vehemently resisted the potential new users, who might or might not adversely affect what they do, but who could surely constrain their right of free access to the OCS. The makeup of the two opposing camps was very interesting. Groups who favored exploration and development of mid-Atlantic offshore oil and gas perceived it as an opportunity for social and economic benefits. These were the oil and chemical industries, heavy construction industries, public utilities, small businesses and skilled-craft industries, highway users, and older, declining cities or towns that wanted to attract industrial development in order to expand and diversify their economic base. Groups who opposed or were wary of offshore drilling included environmental organizations, commercial and sport fishing groups, open-space advocates, commercial resort operators, and representatives of suburban and rural communities.

The two groups listed show that on both sides of the controversy people found themselves sharing concerns with individuals and groups that under other circumstances would have made them adversaries.

Several ironies emerged, centering on the speed and the certainty with which various interest groups perceived the changes that the first proposed Atlantic lease sale represented. There are only a few things that can be said with any certainty about looking for oil and gas in a patently wildcat area such as the Atlantic OCS. First, and foremost, is the simple historical fact that exploratory drilling is going to be unsuccessful about 90 percent of the time. The second is that even after successful exploration, the rate at which development and production takes place is measured in years and decades. Finally, the more highly developed a region is politically, socially, and economically, the more it retains its values, even when subjected to stresses that would completely revolutionize less-developed areas.

If the preceding absolutes about offshore development describe how things generally happen, how did the government's new offshore leasing program in the mid-1970s convince people of exactly the opposite? Part of the answer comes to light if we examine some excerpts from the court testimony given in 1976 by Dr. John Teal, an expert witness in the lawsuit filed by the state of New York against the Department of the Interior in 1975 over the first Atlantic sale, OCS Lease Sale 40:

THE WITNESS (Teal): . . . There is one other point, I think, which is of considerable significance. That is, that the phytoplankton production in the ocean is very much associated with the patchiness of the population. They are not uniformly distributed. In fact, the population of commercial fish and the commercial fishing is generally supported by only maybe one out of every five year's or one out of every ten year's production. That is, most of the years, the fish production does not succeed in producing a harvestable crop. But once out of five or ten years it is successful. And that supports the fisheries for the next five or ten years.

It's generally believed by marine biologists--and there is increasing amounts of evidence to support the idea--that those successful years depend on the association of fish larvae with patches of food supply. The fish, when they hatch, many commercial marine species have very little in the way of food reserves. That is, if they don't hatch in a place where food is immediately available, they fail to survive.

THE COURT: [A] patch would have been as small as this whole area?

THE WITNESS: The patches are of the order of a kilometer or two in size. Based on the--based on studies done in the Gulf of Maine of the size of patches of phytoplankton, the food patches are of that order of size.

THE COURT: Well, it would be many--

THE WITNESS: A few kilometers across.

THE COURT: That would affect the success of a hatch of a whole species of fish?

THE WITNESS: That's what I am suggesting. That the chance occurrence of a patch of fish, larvae with patches of suitable food, might be one that--I have no idea. But that's the--the hypothesis at the present time is that the co-occurrence of the patches of food--and patches of food are what leads to a successful year class in the fisheries.

THE COURT: You're talking about patches of a few kilometers in size?

THE WITNESS: Nobody knows for certain what kind of patches we are talking about in fish larvae. I am extrapolating from known patch sizes of the food, of the phytoplankton production.

THE COURT: We are talking about hundreds and hundreds of miles of fish, aren't we?

THE WITNESS: Yes. But we are talking about the time they hatch out as extremely small animals.

THE COURT: So that just as a small leakage of oil at a critical time could destroy a whole fish hatch.

THE WITNESS: It's entirely possible the patches are about the same size [as] a large oil spill. Now, if you have a large oil spill that coincided with what would otherwise be a successful co-occurrence of fish larvae and phytoplankton, it would be entirely possible to--through several mechanisms, the larvae might be inhibited [from] feeding for a short period, but long enough so that they wouldn't be able to survive. The food--the nature of the phytoplankton may be changed by the sort of thing that I described so that phytoplankton presence would be unsuitable as fish food. So that the survival of the fish larvae would not take place. You would not get a successful year class.

It isn't something that anyone would ever know about except you would apparently just have another unsuccessful year.

THE COURT: What are the probabilities--are these very, very remote possibilities or substantial probabilities that we are talking about? The oil spill itself, in terms of time, is improbable at any specific place.

THE WITNESS: You would have--the oil spill is improbable. There isn't any way I can address that question.

THE COURT: You're talking about--

THE WITNESS: The probabilities.

THE COURT: About a probability, as a substantial effect, [of] 1 in 100 million. It's awfully hard for the Secretary of Interior to consider that in major decision-making processes.

THE WITNESS: My point is that it is not addressed in the [environmental] impact statement. So he doesn't--it was not even presented to him as something which he might consider, and that he--but I have to admit that we don't have a very strong basis for attaining a probability of the occurrence. [New York et al. v. Thomas S. Kleppe et al.]

We can see from the preceding dialogue that even the most knowledgeable people dealing with subjects they know best admit that the technical substance of environmental issues is controlled by a complex series of events, many of which are beyond human control. They also admit that what *can*

happen is often highly improbable and probably would be ex-
cluded as a determining factor in making a decision.

OFFSHORE LEASING AND DRILLING

We can compare federal oil and gas leasing on the OCS to
federal leasing of western coal, another accelerated program, to get a better idea of how the environmental concerns
of coal and oil leasing are related in rhetoric and diver-
gent in fact.

A federal coal lease differs from an OCS oil and gas
lease because, in part, a coal lease is a commodity contract. While an OCS lease gives the right to look for oil
and gas in a particular place, it does not guarantee that
oil and gas exist. A federal coal lease confers the right
to develop and produce an established quantity of coal. The
major environmental difference between an OCS lease and a
coal lease is that as a result of an OCS lease there is a
possibility that adverse impacts may occur some of the time,
but with a coal lease there is a *certainty* that stresses
will be placed on the environment. Because that is the
case, federal management of the coal program requires a much
more thorough examination of effects before leasing can occur. In addition, state governments share revenues generated from the coal program because of disruption of existing
services and demand for new ones.

Similar environmental concerns are possible in all OCS
leases, but probable only in a small percentage of them. An
OCS lease gives the right to drill in an area that may not
contain oil or gas. Therefore, the first risk from offshore
drilling is a financial risk borne by the lessees. Among
the hundreds or thousands of geological structures that may
be explored offshore, perhaps ten may prove to be productive. In a very large province, only a hundred structures
might end up productive.

The exploration phase of drilling operations for offshore
oil and gas occurs on a very small scale. One or two dry
wells can, and usually do, disqualify individual tracts, and
sometimes entire structures, from further exploration. The
amount and nature of such materials as drilling muds and
cuttings, and formation waters are not cause for significant
environmental concern. The major result of exploratory
drilling is the dilution of drilling muds, which are composed of clays and other chemicals, some of which are relatively nontoxic and readily dispersed, particularly in open
ocean currents. However, the cuttings--coarse sands or fine

gravel--bury the ocean bottom beneath the drilling rig. Even so, in time the bottom recolonizes with new organisms.

While there is a remote possibility that even this low level of activity could cause severe environmental concern, the environment would have to be so confined, the life forms so rare, and the conditions so sensitive to all types of change that the probability of this occurring is slight enough to defy description. The government has neither leased, proposed to lease, nor discovered any areas that would be so sensitive that exploration could not take place.

Rather than treat each lease sale or every exploratory well as if it represented a significant threat, we should consider the early stages of *successful* exploratory drilling as the opportunity to establish the number of areas, and their location and size, which are of real concern. The location and size will determine what, if any, resources are subject to harm. We can also quantify the potential for harm and for interference with other activities at this stage. Finally, this would be the time when meaningful and effective environmental restrictions should be imposed, if they are required.

The problem with starting exploration off the Atlantic Coast is not that it has been handled in an environmentally insensitive manner. It is because there has been such a furor over offshore oil and gas operations and there have been so many premature yet unrealized predictions of conflict, damage, and destruction that, if and when problems present themselves in the future, there is a danger that no one will care anymore.

It would be unfortunate if we got to a point in the future where mitigating restrictions were actually required to protect the environment and we were no longer interested in scrutinizing them to make sure they are effective because we had chased too many phantoms in the past.

REFERENCES

State of New York, et al., v. *Thomas S. Kleppe, Secretary of the Interior, et al.* (U.S. District Court, Eastern District of New York, 76-C1229, 76-C208, pp. 352-361)

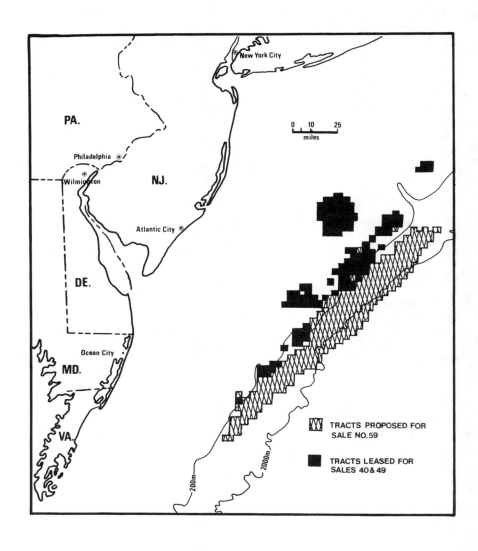

Tracts Selected For Study in Sale Number 59
Environmental Statement

6

Environmental Concerns for Mid-Atlantic Oil and Gas

Bert Brun

The mid-Atlantic bight is a strange amalgam of man's hectic untidiness and nature's forgiving beauty. In no other American coastal region have so many people been squeezed together, directing their inevitable effluvia toward a body of receiving waters. Amazingly, the shorelines remain largely intact, habitat to sizeable populations of living creatures, including nearly 21 million (Bureau of Land Management, 1980, p. 90) humans respite from their bustling megalopolis.

When the mid-Atlantic bight was designated as a frontier area to be leased in 1976 for the exploration of oil and gas on the outer continental shelf (OCS), environmentalists were apprehensive. Images from the Santa Barbara blowout were still fresh; a relentless invasion of gooey oil upon the beaches of New Jersey and Long Island would be even worse than the California experience because of the greater number of people who would be denied precious coastal usage. The question in the minds of many was, Can a natural system already used in so many conflicting ways absorb still another impact without breaking down? Put another way, Would oil or gas production in the mid-Atlantic outer continental shelf be the straw to break the environmental camel's back?

POLLUTION VS. PRESERVATION

Map 4, prepared as part of the Sale 59 Draft Environmental Impact Statement for the Bureau of Land Management's New York OCS Office, offers a good insight into the mix of uses and activities that take place in the mid-Atlantic bight. The map, depicting coastal reaches from Long Island to the entrance to Chesapeake Bay, is marred with such notorious dumping grounds as the "Acid Grounds" where wastes from New York factories are dumped off Sandy Hook, New Jersey; other industrial waste sites where materials even DuPont can no longer use or tolerate are released to the ocean; sewage sludge sites (slowly being phased out) off Delaware and New York Bays; etc. At the same time, an active shipping industry makes its way to and from three major port areas--New York Harbor, Delaware River, and the Chesapeake Bay centers. Ironically, in 1977, the first two industrial complexes featured oil-related processes, using 553 (Bureau of Land Management, 1980, p. 92) million barrels (about 96 percent) of largely foreign oil tankered into the country. One of the arguments justifying exploration and production on the mid-Atlantic shelf has been that if oil or gas are found, tanker traffic, with its inherent risks of spills due to collision or grounding, and the nasty habit of release of oil from ballast tank cleaning, would be reduced and replaced by environmentally preferable pipelines leading directly to refineries. The larger national policy desire to reduce dependence on foreign oil overrides all else, and has been the major underlying impetus toward exploration of frontier areas.

An impression that could be drawn after viewing the pollutant-strewn seascape outlined and described in the BLM publication is that survival appears difficult for all living creatures in these mid-Atlantic coastal waters. Another look at the map provides evidence for nature's outstanding regenerative capabilities: approximately thirty-seven state or federally protected natural areas are located directly on the coastline or behind barrier islands, with others inside the entrances to the great estuaries like the Delaware and Chesapeake Bays. These areas include recreational beach parks, national seashores, refuges and various wildlife management areas, and receive heavy use from nearby population centers. The marine waters of the mid-Atlantic bight are largely responsible for the ongoing vitality of the coastal biological resources. To understand why the region's living resources are important and how they function, the following section gives a brief overview of the mid-Atlantic environment.

THE MID-ATLANTIC ENVIRONMENT

The mid-Atlantic bight is a term of convenience for the shallow indentation in the U.S. Atlantic Ocean coastline bounded roughly by Cape Cod at the north and Cape Hatteras to the south. The area can be sub-divided into two smaller bights, the New York and the Chesapeake. For the purposes of this discussion, and because all the lease blocks involved in BLM's Sales 40, 49, and 59 are located off the New Jersey-Delaware Peninsula coastline, the bight is considered as the waters extending from Block Island, Rhode Island, to Cape Charles, Virginia, and from the shore to the 2,500-meter isobath depth line.

The width of the continental shelf in this region varies from 14 miles off Cape Hatteras to 118 miles off New York (Bureau of Land Management, 1980, p. 65). The shelf is usually thought of as extending from mean low tide to about the 600 foot (100 fathom; some use the 200 meter, slightly deeper) depth contour, or the region where the sharper gradients of the continental slope begin. The slope then plunges fairly quickly to the more downward-sloping continental rise, a kind of sedimentary apron that eventually meets the abyssal plains of the deep ocean.

Geology

The gently sloping mid-Atlantic continental shelf appears relatively featureless and monotonous, for the most part. Structurally it is underlain and dominated by the Baltimore Canyon Trough, a 300-mile structural depression in the crystalline basement rock caused originally by faulting in the late Triassic times. Subsequent continental erosion caused the trough to fill with sediment. The later events of the Pleistocene and Holocene resulted in the existing surficial morphology of the shelf. Advance and retreat of glacial ice sheets, associated with major river flows that deepened into canyon features, left the topmost sediments now present. These canyons are patterned into two major types of features: subsurface, sediment-filled channels flowing roughly outward from old river systems toward submarine canyons at the edge of the shelf, and northeast-southwest trending surface ridge-swale systems largely influenced by oceanographic processes.

The margin between the shelf and continental slope is featured by deep indentations: the submarine canyons that are the remnants of the old river systems. In addition to Hudson Canyon, the deepest and longest of these, at least six other major canyons are present in this region. The slope is underlain by a carbonate reef-platform complex that begins in the northern Gulf of Mexico and extends along the entire Atlantic continental margin. The sediments overlying this reef complex consist of terrigenous (land-originating) materials brought down during late Cretaceous-early Tertiary times. Intermingling of these materials with the reef and with shales and other seaward deposits has apparently resulted in conditions favorable to formation of structural petroleum-bearing traps. It is in this Menard Reef region that the current interest is strongest, in terms of exploration for economically developable finds of hydocarbons. Earlier drilling of sediments in shallower shelf waters turned up little of interest or value.

Hydrology

Three different classes of water are present in the mid-Atlantic bight, bounded by a fourth--ocean water--to the east. Coastal-shelf water, the first class of water, is influenced by contributions from rivers, that can lower salinities to 30 parts per thousand (ppt) or less and that also provide nutrients that enhance overall productivity of the waters. The relative shallowness of coastal waters also creates more pronounced daily and seasonal temperature effects than in deeper waters. However, both coastal and slope waters, the second class, form daily and seasonal (deeper) thermoclines, in which the upper water column is heated to a greater degree than lower waters and is separated from them by a steep temperature gradient. Usually, in the autumn, with the cooling and sinking of upper waters, aided by stronger wind development, an overturn of waters occurs, bringing to the surface colder, more nutrient-laden waters. Slope waters can be subdivided into surface, intermediate, and mid-depth categories, depending on temperature and salinity. Below these, the coldest, saltiest (35 ppt or more) North Atlantic Central Water seldom varies in its characteristics. Bounding slope water and effectively separating bight waters from ocean water, is the third class, Gulf Stream water. Warmer and saltier than slope waters, Gulf waters moderate them; the slope waters thus serve as a mixed water mass at times affected by and at other times affecting the more seasonally-variable shelf waters.

Surface circulation in the bight is generally south or southwestward in direction, out of the shelf break. The northern cell, as far south as Hudson Canyon, tends to receive more indrafts of offshore water than does the southern cell, which is particularly affected by salinity-reducing flow from coastal rivers. Little is known about surface circulation in slope waters, though the shoreline margin (i.e., the area near the shelf break) is thought to contain southward flow currents that turn seaward off Cape Hatteras (this could represent the western side of a cyclonic or counter clockwise gyre or circular pattern). The slope water seems to be frequently invaded by the clockwise-flowing, warm core rings or eddies meandering through the cooler slope waters for up to a year before breaking up or rejoining the northeastward-flowing Gulf Stream from which they originally broke away.

Subsurface and bottom circulation is even less known than that of the surface, in the mid-Atlantic bight. Though this flow is variable, there is a general westerly or southerly directional flow, at least out to the 60-70 meter isobath. The current shoreward of this line also possesses an inshore component that converges on the New Jersey coast, north of 38° 30' N. latitude. This could be of consequence in transporting materials from offshore drilling toward shore. Seaward of the 70 meter isobath, out to the shelf edge, the bottom drift has an offshore component, beyond which little else is known. Beyond the shelf edge (i.e., on the slope), bottom circulation is to the southwest, with a strong westerly component.

Meteorological Conditions

The reputation of the North Atlantic Ocean as a dynamic, storm-ridden system necessitated thorough evaluation of climatic data prior to a decision to drill beneath its deep waters. If weather conditions were so difficult as to make the operations hazardous to man and environment, then extreme caution would be required at every step. The North Sea experience had already done much to confirm the triumph of men and technology over uncooperative marine systems.

The winds in the mid-Atlantic region are mostly from the west, with mean northwest vectors of 8 to 10 knots in January and mean southwest vectors of 4 to 5 knots in July. These prevailing westerly winds are of great importance in moving potential pollutants away from continental shorelines in the event of a spill or dump. Northeastern storms are an exception to these wind patterns; "nor'easters" occurring principally between October and April and generating winds up to hurricane force could push materials in mid-Atlantic waters landward. The high energy of such storm systems as nor'easters, or tropical cyclones (hurricanes), which occur principally from June through November, tends to break up the heavy consistency of oils and to drive aromatic (more toxic) materials into the atmosphere, however.

Winds in excess of 41 knots occur between 0.5 percent and 0.7 percent of the time in the nearshore portion of the mid-Atlantic bight, and up to 1.8 percent of the time in the more offshore areas. A comparison of maximum sustained wind forces shows that the mid-Atlantic curve falls on the less frequent occurrence side compared to those plotted for the North Sea and Gulf of Alaska. Waves generated largely by the winds are a median height of 4 feet in winter; although the mid-Atlantic region is not one of noticeably high waves, they are not precluded--a 44-foot high wave is predicted to occur every two years. Wave heights are a principal factor in the design of platform structures, for example, waves occurring in a "100-year storm" would require a safety factor such that there is a 93 percent chance a platform will survive without collapse over a field development of 30 years (in 1980 a platform did collapse during a North Sea storm, with considerable loss of life). Some data are available for time lost due to severe weather during mid-Atlantic exploratory activities: Baltimore Canyon COST well B-2 drilled in winter conditions (early 1976) and experiencing seas of 55 feet and 60-knot winds, lost 32.5 hours (1.28 percent) of its total time of 107 days on location.

Temperatures are not liable to be a major factor during operations in the mid-Atlantic. The moderation exerted by the ocean reduces January air temperatures of freezing or below to less than 20 percent of the time. Some icing and snowfall do occur, the latter classified only as a "minor" restriction to visibility. Fog, occurring mainly in winter, can limit visibility to less than 500 yards on an annual frequency of 4 to 12 percent of the days (or portions thereof). Fog restrictions would tend to be more hazardous to helicopter or ship traffic than to production operations. Icebergs are extremely rare this far south in the Atlantic (Bureau of Land Management, 1978, p. 439).

Offshore Living Resources

Essentially, the waters of the mid-Atlantic bight are without boundaries, unless the different water masses are considered in this light. Even the water masses, given the same climate and energy influences to surface waters, respond similarly to one another. Because the entire system is so joined and free flowing, a logical approach to describing its living resources is to proceed on an "energy transformation" basis, tracing inorganic substances and physical processes into organic matter, then up a food chain.

The phytoplankton form the vast base of the ocean food chain and also influence atmospheric conditions, such as by generating great volumes of oxygen. In taking up forms of nitrogen and phorphorus supplied either from land-based, river-borne sources in coastal shelf waters, or from overturn processes, particularly in the autumn, in oceanic waters, the phytoplankton generate the carbon sources needed to sustain higher marine life. The familiar photosynthetic process, using the sun's energy and dissolved gaseous oxygen and carbon dioxide, drives this buildup of organic matter. Phytoplankton in the mid-Atlantic are not well-known, but some generalized data are available. Species composition is frequently important, both as indicators of resource to environmental conditions by the phytoplankton and as influencing the food chains, based on different species.

In the open ocean, the species of importance are the coccolithophores and dinoflagellates. The total number of cells is low compared to the productivity and biomass of the shelf waters. On the shelf, diatoms dominate the species list, while in more turbid, excess nutrient-laden bays and estuaries, chlorophytes sometimes dominate in warm summer conditions; dinoflagellates can also proliferate in warm, shallow, high-nutrient waters, with diatoms more typical of cooler periods.

The typical pattern of phytoplankton production sees a spring bloom instigated by rising water temperatures, increased sunlight, and built up nutrient regeneration. During summer months, nutrients may become limited and the minute plants are heavily grazed. In the fall, a secondary bloom often occurs, chiefly associated with overturn breaking through thermally stratified waters to provide more nutrients.

Zooplankton graze the phytoplankton and are usually the next link upward in the food chain. Like their plant food sources, the zooplankton are limited as to species by temperature and salinity, and certain zooplankton favor or select certain phytoplankton to graze upon. The zooplankton can be divided into two major groups, holoplankton such as copepods which remain as drifters (plankters) during their entire life history, and meroplankton, which exist as part of the overall zooplankton for only a portion of their life cycle. Examples of the latter are larval stages of bivalves and crabs, and ichthyoplankton, which later develop into free-swimming fish.

Slope waters contain only about one-third the standing crop of zooplankton that are present in shelf waters. In turn slope waters have three or four times the crop of zooplankton of the relatively unproductive Gulf Stream waters or Sargasso Sea (oceanic waters to the southeast of the mid-Atlantic bight). However, at the shoreward edge of the slope, that is, the shelf-slope boundary, the waters can be rich in zooplankton. This area is also noted for biological activity by other groups; see below.

The waters of the mid-Atlantic outer continental shelf were intensively studied for zooplankton as part of the BLM's Environmental Studies Program, 1975-77. Virginia Institute of Marine Sciences researchers found three species of zooplankton abundant in the shelf waters--the copepods *Centropages typicus*, *Calanus finmarchius* and the arrow worm *Sagitta elegans*. The latter two species are northern, and tend to drift south with prevailing currents. Maximum biomass was in the spring and mainly inshore (summer and fall), more central in winter and spring.

The inner shelf waters, represented by the New York bight and studies done there, tend to have large copepod populations, dominated by *Acartia* and *Eurytemora* species. Predation upon zooplankton by the next higher trophic level, the secondary consumers (small marine carnivores), can frequently limit their numbers, especially in the "crowded" estuarine or nearshore environment. Zooplankton themselves are limited by their phytoplankton food source, and usually exhibit a lag in abundance after reducing phytoplankton stocks, that is, zooplankton peaks alternate with phytoplankton blooms.

As is typical of oceanic food webs, a wide variety of organisms exists in the mid-Atlantic bight waters, using the plankton populations as a basic food source. These can be grouped in various ways, such as benthic versus pelagic, prey-predator relationships, and resident versus migratory species. For the purposes of this discussion, the biological resources will be separated on a commercial versus noncommercial basis and will include references to other descriptive classifications. Also to be considered as a special grouping are endangered species.

Commercial fisheries in the entire mid-Atlantic area (Cape Hatteras, N.C., to Montauk Point, N.Y.) yielded landings worth a total $92,433,000 in 1979 (Bureau of Land Management, 1978, p. 82). Lobster is the highest valued species and is taken primarily in deeper bight waters in trawls and pots. Lobsters are primarily scavengers and are often found in canyon areas, that is, near the shelf break. They can migrate short distances. Other crustaceans of small, but growing, commercial value include red crab and two other crab species.

Several types of mollusks are taken commercially in the bight. These include surf clam and sea scallop in the shallower waters, ocean quahogs and squid (not recognized by most laymen as a mollusk) in the deeper waters. Bivalve mollusks are typically benthic and sedentary in habit (scallops are mobile). Part of their lifecycle involves a planktonic phase (as does the crustacean lifecycle), followed by the well-known adult burrowing habit. Different burrowers favor different types of sediments. The bivalve mollusks are filter feeders, using plankton/detritus with bacterial burdens as sustenance, and are preyed upon by starfish, crustaceans, finfish, and man (using dredges of various types). Squid are predators, and two species are taken by man in trawls and by "jigging" (by foreign fishermen, who are generally more interested in squid than Americans).

The finfish segment of the fishery represents the biggest share of landed weights, but tend not to be as high in value as shellfish, which are 29 percent higher. For example, menhaden landings in the southern part of the bight, where they are abundant, totaled 732 million pounds (figure includes North Carolina landings) in 1976, worth only $22 million (Bureau of Land Management, 1978, p. 195). Menhaden are an atypically short food chain species, consuming phytoplankton directly. Most other finfish are predators, taking a wide variety of prey, from zooplankton to larger crustaceans, worms, mollusks, and so on, on up to other fishes. Usually the size of the predator determines type of prey taken.

The finfish of most importance in the bight include summer flounder, butterfish, croaker, scup, and striped bass in shallower waters, and black sea bass, whiting, Atlantic cod, mackerel, and two other flounders in deeper waters. Another deep water commercial species, the tilefish, has a unique burrowing habit, especially near or in submerged canyons. Location of finfish tends to vary by season; this patchy distribution, which even applies to shellfish to some extent, makes prediction and commercial exploitation of those resources somewhat difficult.

Generally, most finfish species are further inshore, or migrating along shore in summer and found further offshore in winter. The purpose of the adult migrations may be feeding or may involve spawning. Most of the ocean species spawn offshore but the young swim or are borne into coastal or estuarine waters that serve as nursery grounds. Another migrational group are the long distance, deep-ocean species such as tuna, swordfish, and marlin present in the far offshore bight waters in the summer and taken by commercial and sport fishermen.

The sport fishery in the bight, aside from the deeper-water species mentioned, tends to remain in nearshore waters, where many of the same species listed above are sought by amateur fishermen. Added to the list, and of high interest as sport fishes, are sea trout, bluefish, drum, and spot. As stated earlier, summer finds many of the species in the shallowest part of their range, accessible to relatively small boats.

In describing fisheries of the mid-Atlantic bight, foreign fishing efforts should not be overlooked. Prior to passage of the Fisheries Conservation and Management Act of 1976, foreign fishing fleets dominated the effort on the outer continental shelf. Now that the United States has jurisdiction to regulate and manage fishery resources out to 200 miles, the foreigners (led by communist-block nations, but also including Japanese, Spanish, and other vessels) and their huge factory fleets have had to accept limitations of species and tonnages caught. Their interest now centers on mackerel, red and silver hake, sea herring, alewife, and sea robin; in addition, they take regulated quotas of several other deepwater species such as winter butterfish, whiting, and tuna (taken with long lines), and other species mixed in their trawls. Otter trawls are the main gear of the foreign fishermen, while smaller versions are used by the Americans, who also use purse seines and such coastal water gear as pound nets and set gill nets.

The relative extent of the discussion devoted to commercial organisms should not induce the reader to believe that only these are of importance in the mid-Atlantic bight. The whole supporting network of prey and interlocked, noncommercial marine fauna must also be appreciated, though their common names, if available, would not be very recognizable. For example, the rich diversity of the swale bottom habitats, typical of the region, has been described via the Virginia Institute of Marine Sciences' baseline study program, 1976-78. Polychaete worms, peracarid crustaceans (amphipods, isopods, and so on), sipunculids, small mollusks in the infauna (below the sediment surface), and echinoderms (brittle stars, urchins, and others), coelenterates (for example, anemones), and others of the epifauna (on the surface) serve as prey items to larger predators. The smaller species also perform certain functions, such as reworking sediments by burrowing or deposit feeding, scavenging or detritus feeding (clearing the bottom), and filter feeding that keep the system healthy and dynamic. Bacterial decomposition must also be included.

At the top of the food chain in the mid-Atlantic bight are two groups of special interest to man, though not in a commercial sense. The special vulnerability to oil spills of pelagic birds has become a major concern. A wide variety of these birds, including several species of gulls (most common sea birds in winter); Wilson's storm petrel (the most abundant summer bird), gannets, Kittiwakes, jaegers, shearwaters, loons, phalaropes, and others are present in the region, depending on season. The shelf-slope break is an important area of concentration for these birds that find good food sources from the upwelled, nutrient-rich waters there building a good phytoplankton food chain base. Another region of concentration, as determined during Rowlett's studies (Rowlett, 1980) for the southern part of the bight, is the 30-40 fathom depth contour area.

The second group of biological resources is endangered whales and marine turtles. The study cited above (Rowlett, 1980, p. 55) also noted a small resident population of fin whales between 30- and 40-fathom depth contours, due east of Ocean City, Maryland. Two small populations of bottlenosed dolphins were also found off the Delmarva Peninsula, one within 10,000 meters of shore, one along the slope contours. Other than these, cetaceans, including the five other endangered whales (blue, sei, humpback, right, and sperm whales) are primarily migrators through bight waters. They proceed northward, fairly close to shore from March through May, and southward, farther offshore, in the late fall. Except for the sperm whale, a large fish and squid eater, the others are baleen whales, which trap plankton and small fishes in large mouth brushes when squeezing out mouthfulls of water. A study just published by BLM's New York Office gives more detail on sightings of these species, other cetaceans (for example, dolphin species), and of marine turtles.

The sea turtles include three endangered species, hawksbill, leatherback, and Atlantic-ridley, and two threatened, loggerhead and green sea turtles. Of these, the loggerhead is the most common visitor to mid-Atlantic bight waters (all are basically more southern based), and is the only one to nest on the beaches of the area. Some of these turtles are omnivorous, others mainly carnivorous, but in every case predatory of a wide range of marine organisms.

Coastal Ecosystems

Coastal waters in the mid-Atlantic bight include some of the best known and most widely enjoyed recreational resources in the world, such as Atlantic City beaches. At the same time, coastal features such as New York Harbor permit a wide variety of water-oriented commercial activities.

The dominant characteristic of the coastal ecosystems in the mid-Atlantic bight is the string of long, low barrier-beach islands and matching spits or peninsulas, behind which are frequently located lagoons and sizeable tidal marshes and meadow systems. Interspersed between the barrier islands are numerous narrow inlets (connections from the inner areas to the ocean), larger embayments, and a few very large bays, which in this region can also be called estuaries because of significant fresh-water inflow. In a few places, such as Atlantic Highlands, New Jersey, or the Palisades along the Hudson River estuary, old crystalline or metamorphic rock complexes directly overlook the marine waters.

Barrier islands such as Fire Island in the northern part of the bight, Long Beach Island in New Jersey, and Assateague Island in Maryland-Virginia are constantly changing in response to oceanic and meteorological forces. The oceanic influence tends to be steady, exemplified by tidal currents and longshore or littoral currents. In the bight, tides are semidiurnal, that is, having two complete cycles every lunar day (24.8 hours). Tides range from 2.0 feet in Montauk Point, Long Island to nearly 6 feet in Delaware Bay. Local geographic configurations have much to do with the height of the tide.

Longshore currents run parallel and adjacent to shorelines. In the mid-Atlantic region, longshore currents can be somewhat variable, depending on seasonal and meteorological conditions, but for the most part their directions are east to west along Long Island's south coast, south to north along most of the New Jersey coast (except for the southern tip, near Cape May) and on Delaware Peninsula to about Ocean City, and north to south below Ocean City. The sediment carried by longshore currents is a source of considerable frustration and anxiety to many resort beach citizens, like those at Ocean City, Maryland, who find their sands constantly being shifted elsewhere. Elaborate beach nourishment schemes and jetty systems have been effected to attempt to compensate or eliminate this littoral drift phenomenon.

Meteorological events can cause dramatic changes to barrier-island complexes. Summer hurricane winds or winter nor'easters often cause inlet locations to change or become

fixed, or even open brand new ones like Ocean City inlet,
opened by a 1933 hurricane.   The storm-tossed ocean waters
can also rise up in fury to wash over much or all of a barrier island, as at Ocean City in March, 1962.   When this
happens, bulkhead or seawall systems tend to fail, and the
unfortunate proclivity of humans to place their structures
too close to the overwash zone reaps its expensive and
tragic reward.

The beach zone of barrier islands is where the normal in-
terchange of ocean force and landform (that is, beach sands)
takes place.   This zone undergoes seasonal changes: mild
summer conditions add sand to the elevated berm area between
tidal effects and interior dunes, while in the winter, more
violent forces tend to erode the berm and shorten the distance to dunes.   The dunes themselves tend to be as fragile
as the tidal area is tough.   Vegetation is vital to the stability of the dunes, helping to anchor them or to reduce
shifts in the face of strong winds and storm-caused overwash.   When development or excessive traffic takes place on
dunes, they tend to weaken or vanish, making the whole barrier island more susceptible to strong natural forces.   In
some parts of the mid-Atlantic barrier island region, extensive development has taken place, a factor that must be considered when additional stress, such as that associated with
OCS energy development, is in the offing.

Behind the barrier islands, which largely protect them
from high ocean energies, are great expanses of mid-Atlantic
tidal wetlands.   Coastal marsh acreage for the five main
states is about 954,000 acres (Bureau of Land Management,
1978, p. 251).   Sheltered from wave action, yet flushed by
tidal circulations entering through inlets and small embayments, these lagoon and marsh regions are highly productive,
biologically.   River and stream input modifies salinity in
such regions, permitting a broad range of organisms that
also benefit from the nutrients and detritus washed down
from land sources, slowed down in the relatively quiet waters behind the islands, and eventually used or deposited in
sediments.

Lagoons and wetlands, like barrier islands, have undergone great changes in recent years.   Originally seen as
mosquito-ridden areas of little interest or value, at first
marshes were grabbed at low cost by industry (for example,
Jersey meadows), for dumpsites or for other industrial uses.
Then their proximity to hard-to-get ocean access became appreciated, and commercial and residential interest became
high, driving up values (an example is the accelerated costs
of wetland property in the casino-thriving Atlantic City

area). Frequently, development for residential usuage such as dead-end boat channels impaired natural circulation, causing stagnation problems. Finally, the ecological value of wetlands came to be better understood and appreciated, and development was slowed by state legislation that limited fill operations usually needed before building. Dredging to maintain channel depths through lagoons and open waters adjoining marshlands is an activity that must be continued, despite the problems caused by ensuing water turbidity and the need for sites for dredge soil disposal.

Coastal Living Resources

The interface between land and sea and between high and lower salinities, a major determinant of species distribution, creates in the mid-Atlantic coastal ecosystem a rich variety of living organisms. The barrier island beaches because of the high energies described above, tend to be a harsh environment for intertidal flora and fauna. Few plants other than diatoms survive. Most animal residents ("psammon," or sand dwellers) are burrowers, equipped to escape crashing waves. Included are several polychaete worm species, various snails and clams, a few crustaceans such as the ghost crab, sand hoppers, and so on. Sand dollars are commonly seen by mid-Atlantic beach residents, as are tiger beetles. To the casual human eye, the most typical and numerous kind of animal of the barrier island is the shorebird. Nearly forty species of sandpipers, gulls, plovers, and other types of shorebirds breed (ten species), winter or migrate through the mid-Atlantic area. The sanderling, probing the wet sand for food items after receding waves, is a typical, familiar example. Some shorebirds are also common in marshes and mud flats. Not to be forgotten as occasional users of sand habitats are marine turtles--the loggerhead is the only species known to (infrequently) lay eggs so far north.

The sand dunes support and are stabilized by several characteristic plants, including sea oats and beachgrass on the seaward side, and seaside goldenrod, poison ivy, beach pea, salt meadow cordgrass, and others, further up the dune. Low shrubs such as beach heather, bearberry, and crowberry may be found on dunes, which when big or protected enough, will also support some woody plants like beach plum, cherries, wild roses, pitch pine, and some deciduous hardwoods such as oak species. American holly becomes more common further south in the region. Using the food and cover

resources offered by dune systems are a particularly diverse
rodent fauna, including various species of shrew, vole, rat,
mice, gopher, rabbit, and squirrel. These are relatively
free from predators except for feral cats and a few raptor
birds such as hawks and owls.

Between the mid-Atlantic barrier islands and the true
shore uplands dwell a diverse, abundant range of organisms
in the estuarine waters, mud flats, and marshes. The relatively still waters (compared to the ocean) that adjoin,
flush, and replenish marshes offer a type of habitat especially favorable to fishes of the mid-Atlantic. Some fish
species are resident (for example, small prey items like
killfish); some spawn in one part of the marsh or estuarine
system, typically fresher waters, then work their way down
to saltier waters for adult life (perch); some spawn in
fresher waters, migrate out to sea by adulthood, to return
later for spawning (true anadromous fish like river herring
and striped bass, a premier sport fish of the mid-Atlantic
region); eels, a catadromous species, reverse this process;
a large group spawns in the ocean, the fry or juveniles become transported into marsh estuarine nursery grounds for
rich feeding on plankton or smaller fish, then leave for the
ocean again as they approach adulthood. This group includes
menhaden and gizzard shad. Another large group, especially
in abundance, consists of those species visiting the rich
summer feeding grounds as adults, then migrating back to the
ocean (for example, bluefish, sharks). In a typical mid-
Atlantic marsh-water habitat 43 species (Bureau of Land Management, 1978, p. 266) were found, many of commercial value,
using the locality in one or a combination of ways listed
above.

In many of the shallow, fringing lagoonal or estuarine
waters, beds of submerged aquatic grasses such as eelgrass
and widgeongrass in high salinity waters and several other
species of brackish waters, serve as particularly good habitat for small fishes, crustaceans, and other small aquatic
creatures needing protection from predation. Two examples
of valuable crustaceans of the grass bed are grass shrimp, a
highly preferred fish food, and soft shell (molting stage)
blue crabs sought during this vulnerable period by many
predators, including man. Man also exploits molluscan resources such as oysters, clams, and bay scallops that dwell
in the shallow water sediments of the region.

Food webs in shallow estuarine waters are usually extremely complex, facilitated by the high levels of nutrients
available for primary production at the base of the food
chain. Some estuaries in the area are threatened by

excessive nutrient levels (from sewage, fertilizer, and so on) and by toxic materials that tend to accumulate in sediments.

Emergent plants tolerant of salt and of regular tidal inundation constitute the typical marsh habitat. The most characteristic emergent marsh grass of the vast acreages in the mid-Atlantic region is salt-marsh cordgrass. Three-square and other *Scirpus* species are also common.

Wetlands are those above average high tide but still subject to intermittent or storm-caused inundation. Black rush, salt meadow grass, and saltgrass are common, more salt-tolerant wetland plants, while reedgrass (*Phragmites*) and cattail tend to favor fresher water situations. As the land becomes drier, shrub or tree species begin to survive.

Terrestrial but aquatically-oriented wildlife abound in marsh habitats. Some feed directly on marsh or wetland vegetation; muskrats and certain waterfowl (in part) feed in this manner. Other animals are predators; raccoons seek small crustaceans such as fiddler crabs, and ospreys and eagles seek fish. Furbearers are well represented and are taken commercially from many mid-Atlantic marshes. These include muskrat, nutria, mink, raccoon, and others. Some deer, foxes, skunks, possums, and many rodents are also present, with all their supporting food web species.

Waterfowl species, already well represented in marsh-wetland habitats by such species as clapper rails, coots, gallinules, sparrow species, red-winged and other black-birds, marsh hawks (in decline), are augmented in summer by various wading birds. Majestic great blue herons that sometimes also winter over, several other heron and egret, for example, snowy and cattle egrets, which are on the increase, are included in this group. Very large numbers of migratory waterfowl choose the mid-Atlantic region from mid-New Jersey southward, and especially the Chesapeake Bay, in which to winter over. The most numerous ducks are greater scaup (85,000 in the Montauk-Cape May portion of the area in 1975) (Bureau of Land Management, 1978, p. 241) and black duck. Of geese species, brant and snow geese are common in winter near the coast, and Canada geese present in very large numbers on Chesapeake Bay where they have taken to foraging for left-over corn in farm fields.

Since waterfowl are such a prominent, appreciated, and heavily used (via hunting) resource in the coastal mid-Atlantic region, their management is of prime concern to both federal and state resource agencies. Waterfowl management is the major justification for the founding and operation of most of the U.S. Fish and Wildlife Service's 25

National Wildlife Refuges scattered along shorelines of the area. Typically these holdings consist largely of marshes and adjoining waters and uplands, where entire communities of organisms can be studied and managed without interference, though certain multiple uses such as controlled hunting and trapping are permitted. Numerous counterpart state management areas are also dotted along the shorelines.

Endangered and threatened species are of special concern to federal and state natural resource managers. The animal species that have been particularly publicized in the region are bald eagles, peregrine falcons, Delmarva fox squirrel, and shortnosed sturgeon. Species management plans are often devised and implemented for such organisms, and when actions are proposed--an example might be an oil or gas pipeline crossing of a marsh--special consultations and precautions must be taken to protect them. In addition to animal species, a list of 24 endangered plants in the mid-Atlantic region has been proposed.

EXPLORATION FOR OIL AND GAS

When Exxon spudded its first mid-Atlantic well on Block 684, in March 1978, the time for discussion of environmental concerns was over. Now the action in the first Atlantic frontier area was on! This did not mean that apprehension concerning the effects of offshore drilling had ended, especially when within three months, a total of seven rigs was busily pricking the mid-Atlantic shelf sediments.

Concern for the environmental effects of OCS development necessarily was directed toward the two types of materials that would, or could, be released into the marine environment, drilling materials and unexpectedly escaping hydrocarbons. Given the nation's predicament that more domestic oil and gas are critically needed, the first action step-- exploratory drilling--had to be accomplished, merely to gain the knowledge of whether or not oil or gas was present under mid-Atlantic sediments in economically recoverable quantities. This decision accepted, environmentalists including fisheries interests, which do not always fit under this general umbrella, nevertheless wanted all possible safeguards taken with respect to the large volumes of drilling materials normally and necessarily discharged during exploratory drilling. Special concern was voiced for the health of ichthyoplankton and other meroplankton (see above), for delicate organisms such as coral, for sessile creatures, subject to burial by drill cuttings, and others.

Therefore, while exploration proceeded apace, much attention was focused in the late 1970s on the fate and effects of drilling discharges. The subject is a difficult one to pin down because of the dynamic qualities of the marine environment. Although everyone acknowledges that the heavier drill cuttings sink more or less directly to the ocean floor under a rig, unavoidably smothering and burying organisms there, the longer-suspended small particles, muds, and so on, are most questionable, owing to the transporting capabilities of ocean currents, and to the potential for great dilution. Even the materials themselves seen as possibly needing dilution, for example, barium, chrome, biocides, all typical components of drill muds, were not completely known as to their effects on living organisms. Therefore the studies organized fell into two general types--transport and disposition of effluents and effects, uptake, and how organisms react.

It is important that studies of drilling discharges be conducted within the same area for which the acquired information is to be cited or applied, because oceanographic conditions can differ widely enough as to make extension elsewhere inadvisable. For example, the current strengths and gyre (circulatory) pattern of the Georges Bank-North Atlantic OCS region are not analogous to the mid-Atlantic, nor are ambient turbidity levels in the Gulf of Mexico (Ayers et al., 1980).

An important and useful rig monitoring study was carried out in the mid-Atlantic during 1979 for the mid-Atlantic Operations Group by their contractor, E.G. and G. Co. Exxon's Block 684, about 97 miles off the New Jersey coast, was the location of the drilling operations carried out by the semi-submersible rig, Gomar Semi-I, in 120 meters of water. Rig monitoring covered pre-, during- and post-drilling periods. During drilling, which went to 4970 meters, the rig discharged 752 metric tons of barite (Barium Sulfate), 1,409 metric tons of other, lower gravity solids, and 94 metric tons of organic chemicals, including chrome lignosulfonate. Approximately 60 to 70 percent of the materials released were borne away by currents, the rest sinking to the bottom immediately. In the study area, prevailing currents were to the south-southwest, and ranged from 5 cm/sec at surface to 22 cm/sec at 10 meters. A gravity factor plays an important part in the formation of two plumes, the lower, more rapidly descending one of heavier materials (southerly vector) the upper plume remaining more diffuse and drifting longer with the current, southwesterly. The upper plume, with much colloidal material also tends to remain above the thermocline.

The only two water quality factors affected by the upper plume were suspended solids and transmittance. Background levels for the former were reached 350-600 meters downcurrent from the source, and transmittance background values by 800-1,000 meters downcurrent. In addition to water quality factors, some localized physical alterations of bottom sediment, such as increased clay particle presence, were noted up to a 2-mile radius from rig site.

Most trace metals in the discharges such as cadmium, mercury, lead, nickel, vanadium, and zinc, were attributable to formation drill solids, that is, taken from the sediments. Presumably these natural materials were in a solid or bound form not easily taken up by organisms. Barium and chromium, since they were added to the effluent (after use in the drilling) showed expected elevations, with some zinc or lead elevations also possible from this source. Since all mid-Atlantic rigs downshunt drill effluents, even the materials in the upper plume would not be expected to come into significant contact with delicate plankton organisms near the surface of the water column. Light needed by phytoplankton would also be most heavily screened below the most photosynthetically active zone.

The effects question was also looked at in the mid-Atlantic study. Three animals, representing different phyla, were analyzed from the standpoint of uptake of trace elements. Indigenous brittle stars, mollusks, and polychaete worms collected during the pre-drilling survey in Block 684 were found to have significantly higher barium and mercury content during post-survey tissue analysis. The barium elevation was predictable, the mercury not, since negligible mercury was discharged. Chromium was elevated only in polychaetes. Other metals (see above) were not elevated beyond original concentrations. Since most of the organisms were taken in fresh condition, lethal effects of barium and mercury are not known from the uptake determined. Longer term monitoring, or possibly laboratory testing of sublethal dosing, would be needed to gain such information.

Another, less ambitious exploratory rig monitoring study was begun in September 1980 at the request of the mid-Atlantic Biological Task Force near Exxon's Block 816. This undertook to ascertain current directions presumed to be toward nearby Toms Canyon, sedimentation rates, and post-drilling barium and chromium concentrations, compared to the pre-drilling levels in sediments. Biological samples were also taken for possible later analysis, if more holes are drilled. Incomplete results to date emphasize, mainly, predictable barium elevations.

To date, no work has been done on the effects of drilling materials on corals in the mid-Atlantic area. Though soft coral groups like sea pens are known to exist on the slope elevations, not too much is known regarding abundance of hard corals; some are believed to exist in some mid-Atlantic canyons. Monitoring and testing of hard, reef-building (unlike mid-Atlantic) corals have been undertaken in the Gulf of Mexico. To the degree that extension can be made of these data, all species of Gulf corals could survive 96 hours within 1 meter of the source of mud disposal, during normal drilling techniques. At 96m from the source, the concentration of suspended solids was 132 times less than that required to cause mortality in 96 hour bioassays. In other words, normal dispersion and dilution seems to give Gulf corals a good chance of survival. In another Gulf field study, highly concentrated doses of unused drill mud did reduce growth of one hard coral, while in a Florida Keys study, of seven coral species, significant mortality occurred within 65 hours for 3 species, at 1000:1 dilution. Polyp retraction and pump malfunction, causing smothering to take place, were mechanisms noted (Thompson and Bright, 1980).

Presumably, mobile species such as fish can avoid the worst areas of turbidity and toxic material content, in the proximity of rigs. Those sessile organisms quite close are likely to be impacted, and passing plankton too may suffer some deleterious effects from rig discharges. The key to the matter seems to be in the degree of dilution that occurs, and to a lesser degree, the physiological effects of materials that are taken up in such limited nearby areas with high concentrations of smothering or toxic substances.

Hydrocarbon Production

Most environmentalists and those concerned with living marine or human use of coastal resources have been willing to concede that the drilling-materials issue is a pale one compared to hydrocarbons. If a large find were to be made, then a conceivable forest of producing rigs, oil-storage pumping stations, and gas-compressing stations could someday fill the mid-Atlantic offshore seascape. A distinction should be made as to gas versus oil being produced by such rigs--if gas is the only commercially viable hydrocarbon found, as most evidence so far indicates may be the case, then the potential for environmental disaster is much reduced. Oil presents far more serious problems when spilled.

The modest medium-find projection, compared with other reserves or producing areas, was an issue in itself prior to the first Sale, No. 40. The question was raised, Why bother for potentially small production? It was decided all potential sources must be explored.

The worst scenario could consist of a large foreign tanker, which still will deliver imported oil for quite some time after near-term new U.S. discoveries, being turned off course by storm or fog or human error, like the Argo Merchant, colliding with a rig, causing both to spew out huge volumes of oil. To lessen chances of this the Coast Guard studied the need for Port Access Routes to be set up in the busy shipping lanes of the mid-Atlantic OCS region. As part of this, a Fairways Scheme by the Council of American Master Marines has been proposed. The Coast Guard concluded that ship separation schemes were not needed at this time.

Another scenario could be the rupture of a pipeline by one of the numerous, heavy, otter-fishing trawls used in the area. Incidence of such rupture does not seem frequent in other areas, however, and pipeline flow control points could possibly handle this before too much is lost. Trawls might be the bigger losers.

Last in the scenarios for spilled oil would be the possibility of a blowout, during production, or even exploratory, drilling. The IXTOC experience in the Gulf of Mexico illustrated the length of time and massive spillage that can occur, even in relatively shallow waters when a blowout occurs. With the interest in the mid-Atlantic now being focused on the deeper waters of the slope, the complications that could arise if a blowout at 1,000 feet or more could be extremely important in terms of controlling the flow. Industry insists that it has the technology to do more than drill in deep water to 4,876 feet, offshore Canada (Bureau of Land Management, 1980, p. 63). Though specialized rigs are few in number, industry provides types of subsurface production systems such as compliant structures in 3,000 feet of water, subsurface safety valves, and so on that safely provide for production. Pipelines have also been laid in 2,200 feet of water.

Nevertheless, there is a finite probability that spills will occur for the mid-Atlantic region, should oil be developed and produced there. Should a major volume of oil be loosed in the area, as opposed to a small spill or minor chronic leakage, the direction it travels will be of crucial environmental importance. As an aid in predicting oil movements, the U.S. Geological Survey has developed a mid-Atlantic Oil Spill Trajectory Analysis Model, using known wind and current information. The model assumes that the oil mass acts as a single point unit that after certain time periods may intersect different targets such as coastlines. As an example, the probability of a spill of more than 1,000 barrels from somewhere in the mid-Atlantic lease sales area hitting land within 3 to 10 days is less than 0.5 percent. From the nearest-to-land tract area, a 3 percent probability occurs of land impact within 30 days. Time elapsed is important, since more breaking up, weathering, and general reduction of the more toxic, volatile fractions in crude oil would occur over more time.

Probable effects--offshore

Whenever an oil spill occurs scientists have an opportunity to study its effects. Although the offshore portions of the mid-Atlantic bight region have thus far been spared a major spill, enough observations have been made elsewhere to be able to describe the probable effects there. The 7.7 million gallon spill of fuel oil from the *Argo Merchant* that broke up off Nantucket Island, Massachusetts, in December 1976, provides the nearest and best-studied analogy for the mid-Atlantic region.

In the *Argo Merchant* case, prevailing seasonal winds and currents fortunately directed spilled material away from shore, so all study results pertain to marine organisms. Little phytoplankton analysis was done; previous reseach has indicated that these organisms quickly reestablished normal population levels after a spill because of rapid reproduction rates and high recruitment capability from unaffected nearby areas. Of the zooplankton community, copepods were significantly contaminated by oil from the *Argo Merchant*. These animals, which are a major food chain item for small fishes of the mid-Atlantic, were both covered externally and had ingested oil particles in and around the slicks. Aside from clear cut smothering of gills, where oil was heaviest, immediate detrimental effects were hard to determine. Ingested oil was often expelled in fecal pellets. Fish eggs, a meroplankton element of the zooplankton community, were heavily affected near the spill, for example, 98 percent of pollock eggs at one station were dead or moribund (National Oceanic and Atmospheric Administration, 1977, p. 13). This particular species, present but not abundant in mid-Atlantic waters, has an apparent oil-adherence affinity for its eggs, which have also demonstrated abnormal embryonic development

when oil-touched. At the same station, near the spill center, 64 percent of cod eggs were similarly affected. Cod
are more numerous in the mid-Atlantic.

Six species of fish larvae were sampled off Massachusetts, of which three--cod, hake, and herring--are also numerous in the mid-Atlantic. Another species sampled, the
sand launce, an important small North Atlantic food fish,
had sharply reduced numbers of larvae near the spill. Externally, little oil contamination was seen in larvae, even
though their normal food items, the copepods, were affected
and longer-term effects may have occurred. A laboratory
test on young cod larvae showed significantly reduced survival on exposure to 100 ppb of similar fuel oil to that of
the *Argo Merchant* (Kuhnhold, 1978, p. 172). Of biota from
37 samples taken, only three adult fish, two cod and one
flounder, proved to have definitely ingested *Argo Merchant*
oil-affected organisms (amphipods).

Benthic samples were also taken around the spill site.
Though easier to sample than plankton or fish because they
are sedentary, they would tend to receive less oily material
except in very shallow situations. There was a little evidence of direct oil contamination, and of all samples, only
two crabs and a starfish were found dead or moribund.

When analysis from pelagic and benthic samples were taken
down to the histological level, various abnormalities, such
as ocular lesions (sand launce) and edematous gills (herring), among others, were found, but could not be traced directly to oil effects. This points up the need for longer-
term, laboratory studies on sub-lethal or chronic effects of
oil on marine organisms; petroleum constituents are thought,
for example, to adversely affect fish fatty acid metabolism
causing cell membrane abnormalities that in turn relate to
the ability to adapt to temperature changes (Environmental
Protection Agency, 1979a, p. 13).

Of the 1121 seabirds observed near the *Argo Merchant* in
December 1976, about 50 percent were estimated as oiled
(Morson, p. 181). Beached-bird surveys on Nantucket and
Martha's Vineyard turned up 181 birds over a two-month period. Of these, nearly half were alcids, though they formed a
small percentage of the birds seen at sea. Alcids, diving
birds, are thought to be more susceptible to oiling than
gulls, the most numerous sea bird. Autopsies of beached
birds showed oil related damage to lungs, kidneys, liver,
and intestinal tract. In view of large populations of potentially very vulnerable sea ducks (eiders and scoters)
normally wintering on waters landward of the spill site it
was especially fortunate that the oil was transported in the

opposite direction. Marine mammals were not shown to be affected by *Argo Merchant* oil.

Although, at least in the *Argo Merchant* case, surprisingly undramatic damage was recorded as occurring to marine organisms, a number of factors must be assessed, such as the type of oil, for example, crude (light, heavy, sweet or sour) or refined, with the latter particularly important as to volatile fractions--normally the most toxic--present. Oceanographic conditions such as sea state, that helps break up oil masses; wind and current directions, so important in the *Argo Merchant* case; and temperature, colder water may increase viscosity or cause more sinking of oil, are of obvious importance.

The area of impact must be viewed as a whole system. It is likely that the zooplankton element will be most adversely affected in the mid-Atlantic just as it was in the North Atlantic; but here season will be very important. Springtime, for example, when most of the finfish species have spawned would be a disastrous time for oil to be loosed. A spill would affect mid-Atlantic herring, or cod, or hake eggs and larvae, and also the copepods on which they feed. While these meroplankton stages appear to be most sensitive to oil, species differences will be important too. Short-term impacts will tend to be more important to young fish or shellfish meroplankton, and longer-term, sub-lethal effects may show up later in larger organisms--the biomagnification effect, after eating smaller organisms. Trophic relationships in the complex mid-Atlantic food web would also play a part, with marine birds or mammals or large fish predators eventually, if not immediately, reacting physiologically to ingestion of oil-contaminated prey. This assumes that their mobility enabled them to escape the immediate physical impacts at the time of the spill. By and large, given the vast space and diluting capabilities of the open mid-Atlantic shelf-slope waters, oil spills tend not to pose the problem that onshore impacts do.

Onshore impacts

Although numerous relatively small spills have occurred in nearshore and coastal waters of the mid-Atlantic bight, the region has been relatively lucky so far. Compare, for example, the truly massive spills on the coasts of Britain (1967, Torrey Canyon, 36 million gallons) and France (1978, Amoco Codiz, 60 million gallons), both of which washed ashore for the most part. One sizable spill occurred in the Chesapeake Bay (killing 31,000 waterfowl) in February 1976, but that estuarine setting is less typical than for ocean systems. The best study of effects on coastal systems, again comes from New England just outside, but analogous to, mid-Atlantic systems. This spill, of 175,000 gallons of No. 2 fuel oil near West Falmouth, Massachusetts, in September 1969, has been intensively studied over a long period (Environmental Protection Agency, 1979b).

The oil spilled into the marshes and sediments of the West Falmouth area has persisted for over ten years, traceable to the gas chromatographic fingerprint of the originally spilled oil. This points up the slow degradation of heavier hydrocarbons, some of which has slowly dissolved into the water, but mostly taken up by organisms such as grasses, algae, and then up the food chain, by mussels, fish, and birds.

The immediate effects of the West Falmouth spill included mass mortalities in benthic communities. Shallow waters enabling the oil to reach bottom, and oscillating tidal currents keeping it local, helped to exert this disastrous short-term impact. Trawls in ten feet of water brought up collections of all sorts of benthic organisms, including several fish species, of which 95 percent were dead or dying as the direct result of heavy oil contact externally, smothering of gills, or internally, causing physiological malfunction.

Recovery has been slow, especially in heavily oiled locations. From a low reached of two to three surviving species, at a heavily oiled spot in 1969, only about 25 had reestablished by early 1972, compared to the 65 at the control (un-oiled) station. Greater fluctuations also were a feature of the long recovery period, as was occasional heavy dominance by certain opportunistic species like the worm *Capitelli*. Sublethal effects must also be considered during long recovery periods at a heavily oil-affected coastal location. In fiddler crabs, for example, behavior patterns changed, molting increased, abnormally shallow burrows led to high winter mortality and sex ratios were disrupted. On the ecosystem level, sublethal effects included other behavioral disruptions for many animals, thermoregulation interference, changed reproduction and growth rates resulting in abnormal population age structure, altered competitive balance, and predator-prey interactions.

In short, heavy oiling in coastal waters, where the oil becomes thoroughly infused into the basic ecosystem structure, that is, the non-living sediments, to which oil molecules can absorb, as well as uptaken into living tissue, can result in long-term changes. This can be predicted with considerable confidence to occur in New Jersey or Delaware marshes just as readily as in Massachusetts, with the following qualifications--the distance offshore of the site of the spill relative to the target landfall site could greatly modify both amount of oil and toxicity of the impacts. This has been a frequently repeated recommendation, by the Fish and Wildlife Service and others, that is, that lease blocks closest to shore be deleted from sales, in order to increase possible travel time before oil hits beaches or marshes. With particular reference to wintering waterfowl, the longer the oil that might be headed ashore has to weather or break up, the less should be the impact on the geese and ducks sheltering in the many management areas dotted along the coastlines of the bight.

Onshore Facilities

To some degree the potential impacts of spilled oil on coastal natural resources can be avoided, or prospects for mitigation improved, by the wise choice of a landfall site. In the mid-Atlantic area the most likely landfall of an oil or gas pipeline is the New Jersey coastline. Both the federal government, with jurisdiction in the offshore zone, and the state of New Jersey, with jurisdiction out to three miles seaward from the coast have prepared documents (Bureau of Land Management, 1981) reviewing factors involved in possible landfall sites and pipeline corridor selection. These references also provide technical background material on pipelines themselves. Coordination is necessary in the marine pipeline selection process for obvious reasons of producing a continuous transport route. On land, the pipeline corridor route is more directly in the hands of the state (primarily, the New Jersey Department of Environmental Protection), although a considerable number of federal regulations, procedures, and permits must also be satisfied (Rutgers University, 1980).

The landfall site chosen in the mid-Atlantic region for a pipeline to come ashore is likely, almost of necessity, to intersect barrier islands and marshlands. Because of the fragile nature of dunes on barrier islands and the intricate web of life in the marshlands, as discussed earlier, utmost care and precaution must be carried out in such pipeline crossings. For example, dune height and sand stabilizers such as snow fences and grass plantings must be maintained; replacement of correct sediment, preservation of hydrological characteristics, and eventual restoration of original vegetation types that will serve as habitat for fauna, must be achieved in aquatic and wetland habitats. Seasonality must be a considered factor in any landfall pipeline laying work, relative to sensitivity for organisms such as breeding or rearing periods.

Many other onshore facilities are involved when hydrocarbons are explored, produced, or transported ashore. Some, like support bases, pipeline fabrication and pipe-coating yards, oil-storage terminals, refineries, and petrochemical industries, can eat up large amounts of land area and issue potentially harmful effluents controlled to some degree by environmental discharge regulations. In the mid-Atlantic region many of these facilities are already in place, for example, the Davisville, R.I., marine support base, Atlantic City heliport, and the refineries of northern New Jersey and upper Delaware Bay. Should a large find occur, the pipeline requirements would have to be met, but not necessarily in sensitive coastal areas, though one company, Brown + Root, did have plans for a fabrication yard in the lower Eastern Shore of Chesapeake Bay in Virginia.

A gas processing plant would necessarily have to be located between the landfall point and the land corridor pipeline union with a main carrier line. The farther inland the better would be the recommendation here to avoid using up coastal habitat, if possible. For a plant processing under 600 million cf/d, about 75 acres would be needed (Conservation Foundation, 1978, p. 205), of which up to 55 to 65 acres is actually buffer and still of some habitat utility.

From the gas processing plant--the most likely possible major facility now looming, in terms of results-to-date in the offshore drilling--the pipeline corridor would have to be chosen, over a route offering as few constraints as possible, and leading to the most likely junction near Princeton, New Jersey. The federal and state publications referred to earlier offer matrices for selection of corridors in the sea as well as on land, in which biological-environmental considerations are taken into account. Upland habitats, though usually less sensitive and easier to work in than coastal ones, must also be excavated with care and restored to original conditions.

The use of existing rights-of-way such as parallel highways or railways, probably provide less interference than most other terrain-crossing options. In the case of New Jersey, its famous Pine Barrens are a particularly unique, beautiful type of habitat through which the most logical pipeline corridor would pass. This issue will have to be resolved and if the Pine Barrens are transversed, rights-of-way will have to be used to best advantage.   Other extreme precautions will have to be taken so as not to unduly disturb faunal resources such as warbler populations and spring-fall migrations and social patterns such as traditional residents' life styles.

PROTECTING THE MID-ATLANTIC ENVIRONMENT

An important part of the federal role in developing offshore oil and gas is the responsibility for understanding and protecting the environment.   The environmental studies program of the Bureau of Land Management has been the principle vehicle through which to accomplish this.   In the mid-Atlantic region these have included, in addition to several purely geological and physical oceanographic studies, the following completed major biologically oriented studies: a benthic survey of two sites in the Baltimore Canyon Trough (National Marine Fisheries Service, or NMFS); an ambitious, two-year study program that inventoried and described virtually all aspects of the living mid-Atlantic marine environment (Virginia Institute of Marine Sciences, or VIMS); and two other NMFS studies that looked at diseases of fin and shellfish in the region and also at ichthyoplankton distribution and abundance from historical data.
   Because of perceived gaps in knowledge, two other subject areas have been studied in recently concluded or still ongoing studies.   Canyon organisms and processes have been the focus of one of these; this is intended to provide information needed to relate to the shift in industry interest in deeper-water, canyon-prevalent portions of the shelf slope. The other study has been to gain additional distributional and abundance information concerning species of marine turtles and mammals, in the mid- and North Atlantic.   Associated studies have looked at the effects of oil and sound on marine mammals.   In addition to the active investigations providing new information, the BLM program, through its New York OCS office, has also funded several comprehensive literature searches and descriptions of the mid-Atlantic environment, using existing information.
   Information gained from the environmental studies program is intended to be most useful in terms of designing or implementing mitigation of potentially harmful effects of the OCS development process.   One way in which this can be accomplished is via Stipulation 2, the Biological Resources Stipulation that allows for surveys of specific lease blocks to determine whether biological species or populations

within them need special consideration. If so, as determined by the USGS deputy conservation manager for offshore field operations, the lease block operations can be moved or modified so as to protect the resource.

In the mid-Atlantic this process is coordinated via the Biological Task Force (BTF), an interagency group set up in 1977 to review environmental information and make appropriate recommendations. One of these was an early recommendation to shunt, or downpipe, the release of drill muds and cuttings well below the ocean surface, to better protect plankton. This is now permanently included as Stipulation 4, for the mid-Atlantic. Another recommendation of the BTF was mentioned above, that of monitoring Block 816.

The makeup of the task force has gradually evolved and expanded. At first it was intended to provide additional deep-water expertise, beyond the purely Department of the Interior provisions for environmental protection. The latter arena was originally shared among several agencies, of which only three--BLM, Geological Survey, and Fish and Wildlife Service--took active part. Of these three, BLM (pre-lease activities and transportation responsibilities) and GS (operations) were necessarily primarily development minded, leaving the bulk of environmental review in principle to FWS. The differing agency tasks were outlined in then Secretarial Order 2974, now Departmental Manual (Part I) Section 655. Along with the NMFS expertise, water quality expertise was seen as needed, so the Environmental Protection Agency was invited to participate. EPA already had a system, the National Pollution Discharge Elimination System (NPDES), through which certain materials like heavy metals and grease were to be limited and monitored in the OCS mid-Atlantic region.

When it was becoming obvious that the mid-Atlantic was finally going to be actively explored for oil and gas, the various states fringing on the waters of the bight decided to become involved, and they too attended meetings of the task force. By 1978, needed biological/environmental expertise and points of view from a wide range of governmental agencies were included, providing opportunity for a comprehensive environmental review during meetings.

The emphasis on interorganizational review and consensus became more full blown when, in 1979, the Department of the Interior initiated the first meeting of the mid-Atlantic Regional Technical Working Group (RTWG), a member component of the National OCS Advisory Board. The group also participates in National Intergovernmental Planning Program for OCS Oil and Gas Leasing, Transportation and Related Facilities.

All of these entities are part of the framework for coordination and consultation as set up by Section 19 of the Outer Continental Shelf Lands Act of 1953, as amended in 1978; Section 26 of the amendment provides further detail as to intergovernmental liasion. The makeup of the mid-Atlantic RTWG attempts to ensure sound environmental consideration by including private or citizen group and environmental group members, in addition to industry and the usual gamut of state and federal agency representatives. One of the issues the RTWG has looked hard at was the safety of deep-water technology during exploratory drilling in the mid-Atlantic region.

CONCLUSION

The mid-Atlantic bight region has been demonstrated to be a relatively healthy and productive one, from its deepest continental shelf and slope waters right up to its beautiful beaches and teeming marshlands. All those concerned with finding and using hydrocarbon resources possibly stored under the waters wish to keep this status quo if it is humanly possible. The environmental watchdogs are thus joined, at least in purpose, by the would-be developers.

It has been shown that a considerable body of review, description, and research has taken place, in an effort to understand, hence to better safeguard, the living resources of the mid-Atlantic Region. These resources have thrived, or co-existed, along with a host of other uses of the same environment. Thus, multiple use, a policy recently becoming emphasized within the Department of the Interior, has in actuality already been in effect for a long time in this particular case. To date, exploration of the OCS has not appeared to unduly stress the existing, normally-functioning biological activities and processes of the region.

Production of hydrocarbons in the mid-Atlantic, should it become a reality, will impose a need for new examination and all possible precautions as regards the environment. Previously a pause/review period had been announced as a probability between exploration and production phases. The framework described above will then be used to enable all the groups and interests involved to discuss the best means to proceed wisely for maximum environmental protection. With a continuation of the private and public attention previously displayed, it is hoped that the mineral resources of the mid-Atlantic OCS can be obtained with the minimum effects possible on the environment. In this event, the nation will be a clear winner by virtue of enlargement of its domestic energy sources.

REFERENCES

Ayers et al. 1980. "An Environmental Study to Assess the Impact of Drilling Discharges in the Mid-Atlantic. Parts I-IV," in *Research on Environmental Fate and Effects of Drilling Fluids and Cuttings, Symposium Proceedings*, vol. 1 ed. (Buena Vista, Florida: U.S.G.S.).

Bureau of Land Management. 1978. New York OCS Office, Final Environmental Impact Statement OCS Sale No. 49. (G.P.O.).

Bureau of Land Management. 1980. New York OCS Office, Draft Environmental Impact Statement OCS Sale No. 59. (New York).

Bureau of Land Management. 1981. Hudson Canyon Transportation Management Plan, (New York, Outer Continental Shelf Office).

Conservation Foundation, The. 1978. "Environmental Planning for Offshore Oil and Gas," vol. 1: Recovery Technology; (Prepared for U.S. Fish and Wildlife Service, OBS-77/12).

Environmental Protection Agency. 1979a. "Oil Spills." Research Summary, EPA 600/8-79-007.

Environmental Protection Agency. 1979b. "A Small Oil Spill at West Falmouth," Decision Series, EPA 699/9-79-007.

Kuhnhold, W.W. 1978. "Impact of the Argo Merchant Oil Spill on Macrobenthic and Pelagic Organisms, in *Proceedings of the American Institute of Biological Sciences Conference on Assessment of Ecological Impacts of Oil Spills.*

Morson, B. "The Argo Merchant Oil Spill: Impacts on Birds and Mammals," AIBS Conference.

National Oceanic and Atmospheric Administration. 1977. The Argo Merchant Oil Spill - A Preliminary Scientific Report, vol. VIII, Summary Fact Sheet (U.S. Department of Commerce).

Rowlett, Richard A. 1980. Observations of Marine Birds and Mammals in the Northern Chesapeake Bight, U.S. Fish and Wildlife Service, Biological Services Program, FWS/OBS-80/04, (G.P.O.).

Rutgers University. 1980. OCS National Gas Pipelines: An Analysis of Routing Issues, Prepared for New York Department of Energy.

Thompson, J.H., and Bright, T.J. 1980. "Effects of an Offshore Drilling Fluid on Selected Corals." in *Research on Environmental Fate and Effects of Drilling Fluids and Cuttings*, vol. 2, ed. (Buena Vista, Florida: U.S.G.S.).

PART IV

THE
INDUSTRY
PERSPECTIVE

7

The Imperative for Offshore Development

O. J. Shirley

INTRODUCTION

From the perspective of the offshore oil industry and many other interests concerned with domestic energy supply, acceleration of the exploration and development of oil and gas resources underlying the outer continental shelf (OCS) must be given a high national priority. The potential oil and gas reserves under the OCS are of such magnitude as to dramatically increase our domestic supplies and the adverse impacts of exploration and development activities have been shown to be modest compared to all currently viable alternatives. Given these facts the proponents of offshore development view with considerable concern the substantial impediments that continue to forestall expeditious and orderly development of these valuable energy resources, particularly so in light of the nation's dependence on OPEC countries for a large fraction of domestic oil supply.

The issues surrounding OCS oil and gas development are complex and include strong political overtones at both the national and state levels as well as more technical considerations. From an industry perspective the paramount issue is the rate of leasing in the OCS, particularly in the high-potential frontier areas such as offshore Alaska that remain virtually untested. Other important considerations relate to the regulatory and permitting framework within which operations must be conducted.

Opponents of OCS development or acceleration of OCS development have raised many other issues related to environmental and socioeconomic impacts, government vs. industry

145

share of OCS revenues, and conflicts with other activities
on the OCS, particularly fishing, as well as questions regarding the adequacy of OCS technology and safety. This
opposition has been led principally from a variety of national and regional environmental organizations, but has
been supported in several important areas by fishing interests, state governments, and various congressmen.

This chapter will represent the perspectives of the proponents of accelerated OCS development, principally those of
the offshore oil industry. As the issues surrounding OCS
oil and gas development have been addressed in a variety of
forums by many capable spokesmen, a reformation of these
arguments by the author is both unneccessary and undesirable. The reader will be best informed by an examination of
the statements of knowledgeable leaders and scientists from
the industry and trade organizations representing industry
viewpoints. An attempt has been made to segregate these
materials into groupings that address specific issues. However, it will be obvious to the reader that the issues are
complexly interwoven. Thus, the remainder of this chapter
will be devoted to extensive quotations and excerpts from
publications, statements, and public testimony that, it is
hoped, capture the perspective of these proponents.

THE RESOURCE AND THE NATIONAL NEED

American Petroleum Institute

"Energy use in the United States in 1980 declined 4 percent
from the previous year--and oil consumption decreased 8 percent below the 1979 level--but we still imported almost 6.8
million barrels a day of foreign oil. The bill for those
1980 oil imports exceeded $78.5 billion. It is estimated
that the imported oil bill for 1981 will remain at least as
high.

"Much of that oil comes from politically unstable areas
of the world, including the Middle East and Africa. Thus,
we not only import large volumes of crude oil and refined
petroleum products, but events beyond our control could--at
any time--interrupt the flow of this energy.

"The United States does not have to accept such energy
insecurity. If it chooses, by 1990, it could reduce its
petroleum imports to a much safer level through increased
domestic energy production and energy conservation. . . .

"According to 1981 estimates by the U.S. Geological

Survey (USGS), the offshore areas of the United States
to the 8,000-foot water depth may contain:

- As much as 43.5 billion barrels of undiscovered
  and recoverable crude oil and condensate; and
- Up to 230.6 trillion cubic feet of undiscovered
  and recoverable natural gas.

"The upper limits of those resources represent a 14-year
supply of crude oil and an 11-year supply of natural gas at
current rates of production in the United States. The USGS
reports that offshore lands may hold as much as 34 percent
of the nation's undiscovered oil and 28 percent of its undiscovered natural gas.

"In using estimates of undiscovered recoverable oil and
gas, it is important to remember that they are only estimates. They are arrived at by comparing known geologic
features of the area being estimated to other, geologically
similar areas of the world, where sufficient drilling has
taken place to prove the extent of oil and gas reserves.
The more geologic information there is available, the
greater the probability that an estimate will be reasonably
accurate.

"This potential oil and gas--if found and produced--could
go a long way toward providing the United States with energy
and toward further reduction in imports. This offshore oil
and gas could help complete long-term transition from conventional fuels (such as oil, gas, coal, and nuclear power)
to alternative and renewable energy sources.

"But the only way to determine if that potential petroleum is really out there is to drill for it. That can only
be done with timely and adequate government leasing policies
and a determined effort by the petroleum industry. . . ."
[American Petroleum Institute, 1981]

"During the 28 years since the OCS Lands Act was passed,
the Department has leased approximately 20 million acres on
which about 8,000 producing wells have yielded more than 5
billion barrels of oil and more than 44 trillion cubic feet
of gas. We may have, however, only scratched the surface of
the vast energy potential that lies beneath the waters of
our OCS.

"Our current economic situation and the increasing tensions in the Middle East highlight in all of our minds the
need to decrease drastically the United States' dependence
on foreign oil. Reliance on foreign oil not only affects
our economy, it is affecting our security and our role as an

international leader. Energy is the linchpin of the American way of life and we must all work together to increase our domestic energy production, or together we will suffer the consequences.

"For the next few decades, choosing not to use more of the resoucres on public lands will mean choosing lower productivity, fewer jobs, and slower economic growth. At this late date, we cannot afford to be asking ourselves whether greater access is needed. The question is how much more of the energy resources on government lands can be found and developed safely--and with full recognition of environmental values.

"To begin answering that question, let's look at prospects for exploration on the Outer Continental Shelf. The Administration hopes to encourage exploration in this area through a schedule that would increase both the amount and quality of offshore territory to be considered for leasing in the next five years.

"The new leasing approach in the proposed schedule keeps our country going in the right direction. If implemented substantially in its present form, the proposed program will provide dramatic new opportunities to find more domestic oil and gas. . . ." [DiBona, 1981]

J. Robinson West, Assistant Secretary-Policy,
Budget and Administration

"In the past ten years, there has been only one commercial discovery in a frontier area--the Beta Field off southern California. We can and must do better. Our current estimates indicate that there are significant amounts of undiscovered hydrocarbon resources on the OCS--up to 43.5 billion barrels of oil and 230.6 trillion cubic feet of gas. In 1980, we imported an average of 6.2 million barrels of oil a day at a total cost of $78.6 billion for the year, up from $4.6 billion in 1972 and $32 billion in 1976. Because OCS oil and gas costs the Nation much less than imports, the economic losses generated by this huge oil import bill can be reduced by finding more hydrocarbons and realizing the significant economic gains from their production.

"The history of the search for hydrocarbons contains many examples of years of fruitless drilling in a region, then suddenly a commercial find, followed by continued successes. Prudhoe Bay in Alaska, the Hibernia field in the Canadian Atlantic, the North Sea, are all prime examples of this scenario. We believe that by broadening the range of possibilities from which industry can select, we will expedite this process in those OCS areas where commercial deposits of hydrocarbons may eventually be found. . . ." [West, 1981]

SAFETY AND THE INVIRONMENT

American Petroleum Institute

"Much of the delay in federal oil and gas leasing, particularly in offshore areas, is the result of unwarranted opposition. In a number of instances, state governments and environmental groups have blocked or delayed lease sales through court challenges. Moreover, environmental laws and regulations--including the Coastal Zone Management Act of 1972, the National Environmental Policy Act, the Clean Air Act and a host of others--have been used to delay offshore operations.

"The history of petroleum operations in the United States clearly demonstrates, however, that oil and gas exploration, development, and production are compatible with environmental protection.

"Through 1980, more than 25,000 wells had been drilled in U.S. waters. Yet there has been only one offshore platform accident which resulted in significant quantities of oil reaching nearby shores. That was the accident in the Santa Barbara Channel in 1969. Both government and private scientific investigators report that there is no evidence that permanent damage resulted from that oil spill and that the area has recovered from any temporary damage.

"In the northern Gulf of Mexico, the most explored offshore geologic province in the world, no significant adverse environmental effects have been reported. . . ." [American Petroleum Institute, 1981]

J. Robinson West

"The overall safety record on the OCS is a good one, and its environmental risk is low compared to other contributors to marine pollution. For example, it is estimated that offshore production is responsible for less than 2% of the petroleum entering the world's oceans, a little over a million barrels worldwide--less than one-fifth of that contributed by natural seeps. Routine tanker operations, such as tanker washings and loading operations, contribute 29% of the oil annually entering the marine environment, or 13 million barrels, and tanker accidents another 5% or 2.3 million barrels. River run-off is responsible for 11.7 million barrels, and municipalities and industrial waste, 8.8 million." [West, 1981]

Charles J. DiBona, President,
American Petroleum Institute

"Actually, there is nothing unbelievable about the world of offshore platforms. These structures simply provide one of the finest forms of artificial reef. They offer both food and protection--the chief survival needs for any form of life. Incidentally, man-made tropical artifical reefs are three to eight times more productive than natural reefs and two to six times more productive than land under intensive cultivation. Low profile reefs are much less productive than high-profile reefs--and the highest profile artificial reefs are oil platforms.

"The environmental benefits to all Americans from petroleum development in the Gulf of Mexico are many and varied. Only a few have been discussed here. It should be emphasized above all, however, that the quality of life in the Gulf Coast, as elsewhere, is inexorably tied not only to a rich environment, but also to adequate supplies of energy. The experience of petroleum development on the Gulf Coast that has been briefly documented suggests that we can have both.

"Dr. Lyle S. St. Amant, Assistant Director of the Louisiana Wildlife & Fisheries Department, confirms this optimistic judgment. He has witnessed three decades of concurrent development of the oil and fishing industries on the Gulf Coast and is one of the most knowledgable experts on the associated benefits and problems. Recently Dr. St. Amant observed that most of the problems which have occurred have been people problems, not environmental ones. As far as we know, in the 30 years I've been involved, and with more than 2,000 oil platforms operating, there has been no real danger from offshore oil drilling. In fact, we haven't found any connection between fish production and oil, except that some of the best production is in the areas where the oil fields are.

"The Gulf Coast does indeed provide living evidence that it is possible for platforms, oil, fish, shrimp, waterfowl, and other wildlife to co-exist in a viable relationship. It demonstrates the petroleum industry's commitment and ability to develop energy resources without irreparably disturbing the environment. And, although not problem free, it also shows that petroleum development can provide sometimes unexpected but very welcomed benefits to man and nature alike . . ." [DiBona, 1981]

Dr. F. T. Weiss, Sr., Staff Environmental Representative, Shell Development Company

"The National Academy of Sciences' Report regarding the biological impact of the Santa Barbara spill concluded that there was no directly attributable damaging effects from the oil spill on large marine mammals or on benthic fauna and the area recovered well within a year.

"A thorough study following the Chevron platform spill in the Gulf of Mexico in 1970 showed that the spill produced no measurable impact on the marine biota.

"A number of other studies have come to fruition in the last few years, addressing such subjects as the environmental impact of routine drilling, drilling muds and cuttings, food availability for plankton, and suffocation of benthic, or bottom-dwelling organisms. Three separately conducted and funded studies which made environmental assessments of petroleum operating areas in the Gulf of Mexico are worthy of note:

"1. The study conducted by Southwest Research Institute, funded by the Bureau of Land Management, on 'Ecological Investigations of Petroleum Production Platforms in the Central Gulf of Mexico';

"2. The study conducted by National Marine Fisheries Service (Galveston), funded by the U.S. Environmental Protection Agency, on 'Environmental Assessment of the Buccaneer Oil and Gas Field off Galveston, Texas';

"3. The study conducted by the Gulf Universities Research Consortium Offshore Ecology Investigation, funded by a group of petroleum companies, examining Timbalier Bay, Louisiana, during the period 1972-1974.

"These three studies--again, all conducted and funded separately--came to essentially the same conclusion: The ecosystems around offshore petroleum operations are normal and healthy; they are not adversely impacted by such operations. . . ." [Weiss, 1979]

Edward W. Mertens, Senior Research Associates, Chevron Research Company, Subsidiary of Standard Oil Company of California

"Under unusual circumstances, marine life can be adversely affected by oil. Oil spills can be toxic, but normally only if three conditions prevail simultaneously; namely, (1) the oil is a refined product, such as a No. 2 fuel oil, spilled into (2) a shallow, confined body of water (such as a small bay) during or immediately preceding (3) a storm or heavy surf conditions so that the underlying bottom or sediments become saturated with the oil. Such an occurrence is relative rare, for example, the West Falmouth, Massachusetts, spill in 1969 and the Baja California (Mexico), spill in 1957. However, in such cases, the bottom life (for example, such as worms, crustacea, and molluscs) can experience heavy mortality. Their populations can be reduced for several years.

"Spills of crude oil or of heavy refined products that occur near shore often cause mortality to birds ranging from low to occasionally severe. Such spills rarely, if ever, are toxic to marine life, but their tarry residues washing ashore often smother the intertidal life, especially in rocky areas, such as barnacles, limpets, and mussels. However if such damage occurs, populations quickly establish themselves to normal levels usually within a few months. Such was the experience at the Santa Barbara spill (1969) and the San Francisco Bay spill (1971). There is no evidence that spills occurring well offshore or on the open ocean cause any measurable damage to marine life regardless of whether the oil is a crude oil or a refined product. . . .

"Where studies have been made on the effects of low level chronic exposure of oil to marine life, there is little evidence that an adverse effect on the organisms inhabiting the local area is observed. Such exposure can result from offshore production operations and marine transportation. . . ." [Mertens, 1977]

James G. Watt, Secretary of the Interior

"The environmental record of the OCS program is outstanding. On the federal Outer Continental Shelf, 18,625 wells have been drilled. Of these, approximately 8,000 wells have actually produced; 4,000 wells are currently producing close to 750,000 barrels of oil and 13 million cubic feet of gas per day. Additionally, between 8,000 and 10,000 wells have been drilled on state marine lands. About half of these wells have actually produced oil and gas.

"The offshore drilling performance is, indeed, dramatically better than that of tankers carrying foreign oil, an alternative source of oil which Outer Continental Shelf production can displace. . . ." [Watt, 1981b]

FISHING AND OIL

Ronald C. Lassiter, President, Zapata Corporation

"During the past eight years, the MIT Sea Grant College Program has .performed a vital academic and public service by focusing on topical issues that affect the world's oceans. None is more important than today's topic: Whether it's reasonable to assume that the Georges Bank can produce both food and energy to meet this nation's needs. This issue is of supreme importance for two reasons:

"One, because of the vast bounty of fin- and shell-fish--and possibly, oil and gas--offered by the Georges Bank area.
"Two, because the debate over the best utilization of the Georges Bank is a micro-cosm of the concerns that separate environmentalists and developers in many areas.

"Much of this debate involves more fiction than function. It is more the product of misunderstanding and misinformation--on both sides--than concrete issues. When the mass media protrays meetings between environmental and energy groups, it often paints a picture of two scarred boxers circling one another warily, watching every move. Frankly, this is too often true. Neither side seems to want to take the time to understand the other's position. And that is sad. If they looked at the facts, they would find that their objectives are not mutually exclusive. I firmly believe that offshore drilling and fishing can coexist peacefully and productively. I make that statement with conviction, and my beliefs are based upon a unique perspective. I'm president of a company that is a major offshore drilling contractor and also has extensive commerical fishing operations. . . .

"In considering the effects of petroleum exploration and development on the fishing industry, we have to consider the situation that has occurred in the Gulf of Mexico. During the past twenty-five years, offshore drilling has grown from nothing to extremely heavy development in some areas, especially offshore Louisiana.

"There are perhaps 3,100 offshore structures located off Louisiana today. During the same twenty-five year period, we have seen consistent increases in the menhaden harvest in the same area, growing from a total of 213,000 metric tons in 1955 to a peak of 820,000 metric tons in 1978. This fact is of key importance, because menhaden are a delicate species, as are shrimp and other shellfish, and depend on a clean estuarine system of survival.

"Unlike sportfishing, the increase in our menhaden harvest is not related to the presence of oil rigs. A larger fleet and better fishing techniques largely account for the increased catches. But the fact remains that fishing, in general, continues to be very good in the Gulf--even in areas of heavy offshore petroleum development. To me, one would have to conclude that the presence of rigs has not been detrimental to the fishing industry.

"The Gulf of Mexico is not the only area where commercial fishing has continued to flourish side-by-side with the oil industry. Look at the fish catches of recent years in the North Sea, offshore West Africa, and even along the U.S. Pacific Coast. Fishing remains rich in all those areas, despite extensive offshore petroleum development. When there is a problem with a particular species it can generally be attributed to overfishing. No causal link to gas and oil exploration and production has ever been proven. . . .

"I cannot foresee a significant spatial problem caused by platforms and temporary rigs on the Georges Bank. My conclusion is supported by a 1976 study by the Woods Hole Oceanographic institution entitled *Effects on Commercial Fishing of Petroleum Development Off the Northeastern United States.* The study focuses on a worst case scenario for development of Georges Bank.

"In that worst case (or best case in the petroleum industry's view), the study estimates that platforms, grouped in clusters, would preempt approximately sixty-two square miles out of the 20,000 square miles of the Georges Bank. That loss of space would result in a reduction in total catch of no more than one-third of one percent--and probably much less. In the Gulf of Mexico, in the North Sea, and elsewhere, platforms and temporary mobile rigs actually aid mariners by serving as navigational aids.

"I have purposely saved for last the hottest issue concerned with offshore oil exploration. This is an issue so laden with emotion, prejudice, and sensationalism that it is sometimes difficult to discuss objectively. I am talking, of course, about oil spills.

"It is not my intention to brush aside the reality of an oil spill. At their worst, oil spills can foul beaches, kill waterfowl, and make fin- and shellfish inedible. Oil spills are disasters, but they are not catastrophies of the magnitude or the frequency believed.

"That opinion is supported by *The Georges Bank Petroleum Study* published here at MIT in 1973. Considering the possible effects of an oil spill on marine larvae, the report says: 'It appears extremely unlikely that a single large spill will have a noticeable effect on the population of an individual species, especially in view of the fact that these species produce many more offspring than the environment can support at adulthood.'

"It is almost a cliché to say that the United States is a hostage of foreign oil, yet it remains true. We must continue to develop--safely and scientifically--our nation's vast energy resources. Georges Bank holds perhaps 900 million barrels of oil and 4.4 trillion cubic feet of natural gas. It is not a question of whether this nation should develop its available energy reserves. Dire necessity makes the question not *if* but *how*.

"That necessity is not an abstract national goal--nor is it an issue far removed from Georges Bank. The fishermen of the Georges Bank are as profoundly affected by the spiraling cost of marine fuel as the motorist is of gasoline. Small fishermen who operate on a tight profit margin are forced to watch helplessly as scarce supplies and high-priced, foreign imported energy push the cost of fuel even higher. We all have an interest in the development of the oil and gas potential of the Georges Bank, just as we all have an interest in the preservation and promotion of fishing on the Georges Bank.

"I maintain that the two can exist as good neighbors both doing their part to meet vital national needs. We need only to have the vision and the understanding to make it a reality." [Lassiter, 1980]

"Thus, if we are looking for significant effects on commercial fisheries, it is to the toxic effects of oil that we should turn, and we should be concerned not with effects on individual specimens but rather with long-term impact on stocks. In the open sea there is no record of adverse effects on commercial species, and indeed the concentrations of petroleum in ocean waters is so low that a threat could not be expected.

"On the shelf, in the more open waters, we have seen that large spills may cause egg and larval mortalities, but again effects at the population level are not evident, while any impact from point sources such as oil platforms is very localized. . . ." [McIntyre]

REGULATIONS

O. J. Shirley, Manager Government & Industry Affairs, Shell Oil Company

"Compared to a reasonable mode, regulations governing oil and gas extraction on the Outer Continental Shelf are clearly excessive, produce little quantitative benefit to society, and serve to impede the development of badly needed domestic energy supplies. Regulations for the OCS have been developed in isolation without consideration of the interrelationship of impacts created by these regulatory activities, other societal activities, and natural events in the marine environment. If the United States is to expeditiously develop the critically important oil and gas resources in the OCS, the regulatory structure governing oil and gas activities must be radically improved. Such improvements can only be obtaned in an objective manner by centralizing regulatory responsibility and by subjecting all regulatory requirements to a quantitative cost-benefit analysis. Much of the needed reform could be accomplished administratively; however, some legislative changes also seem necessary.

"Data for this report were obtained in a survey of seven companies represented on the Executive Committee of the Offshore Operators Committee, which collectively operate approximately 55% of the active producing wells in the OCS. Data from these companies were plotted (man-years effort vs. wells operated) and extrapolated to obtain an estimate of the total regulatory compliance effort for the entire OCS. Basic findings of this survey are as follows:

- Some 3200 man years of effort valued at $155 million were expended by industry during 1979 to obtain compliance to OCS regulations.
- This level of effort represents about 26% of the operators work force engaged in OCS drilling, development, and production activities. In perspective, 3200 man years/year is equivalent to the primary OCS work force of Exxon, Gulf, Shell, and Texaco which collectively operate 33% of the active wells on the OCS and collectively drilled 22% of all new OCS wells during 1979.

- Stated another way, regulatory compliance effort requires 4-1/2 full-time employees for each drilling rig and 6 full-time employees for each 20-well production platform.
- Of this total effort over half (56%) is estimated to be incremental effort excessive of the industry effort which would be required to operate prudently in the absence of regulation.

"Comparison of regulatory compliance effort in the OCS to other regulatory compliance shows that:

- OCS regulatory compliance effort is about six times greater and nines times more costly than for similar activities conducted onshore.
- OCS regulatory compliance effort is approximately *40 times* more costly than the average for 48 companies (20 industries) participating in the Business Roundtable Study for calendar year 1977.

"A limited effort to identify societal benefits obtained from this intensive regulatory effort reveals the following:

- Oil spilled from OCS drilling and producing operations during the period 1972--1978 (about 1,150 barrels out of the total annual spillage from all sources in U.S. waters of about 375,000 barrels) has improved an average for this effort of 8 barrels per year over the period. Incremental industry personnel costs for regulatory compliance effort is approximately $56 million per year.
- Comparisons between drilling operations conducted in state-owned waters offshore Louisiana and those conducted in the OCS show fewer blowouts per 100 new wells started in state waters which are governed by less severe regulations than in the stringently regulated OCS. . . .

"In the total array of regulations affecting OCS operation there has been no formal effort to determine the benefit to be derived from an imposed regulatory requirement. No effort has been made to illustrate or quantify how the individual elements of a regulatory action will attain the generic benefit. For example, in the area of reduction of oil spills, data show that the amount of oil spilled from OCS operations in U.S. waters accounts for approximately .3

of 1% oil spilled into the marine environment. The societal benefit of regulations designed to reduce oil spills in the OCS, if any, would have a maximum possible benefit were regulations 100 percent successful in removing this relatively small quantity of oil from the marine environment. No government studies have been made as to the actual benefits that would be obtained by society if all of the oil spilled by OCS operations were to be prevented from entering the marine environment. In short, what benefit is obtained by preventing one barrel or a thousand barrels of oil from entering the marine environment? . . .

"There has been a great tendency to regulate the Outer Continental Shelf drilling and production activities in isolation from other industrial and recreational activities that occur on the Outer Continental Shelf. Environmental concerns raised concerning OCS activities include the discharge of drilling mud and cuttings, the discharge of produced water, oil spills, sewage effluence, and the impact of power source emissions and drilling and production operations upon the air quality of adjacent states. Concerns have also been expressed about the chronic effect of these discharges upon the marine ecosystem. Paradoxically, the discharge from outboard motors which operate in the OCS and in more sensitive estuarine and bay areas is exempt from regulation despite the fact that EPA studies show an annual discharge from outboard motors ranging from 220,000 to 670,000 barrels of oil. Sewage outfalls from fishing camps and, until recently, pleasure vessels were exempt from regulation, or such regulations were poorly enforced. No consideration has been given to regulating emissions from diesel engines of thousands of fishing vessels on the OCS which could be a source of chronic water pollution as well as air pollution, and which collectively have a combined horsepower rating exceeding by several orders of magnitude the power plants used in OCS operations.

"Despite the high standard set for industrial activities by EPA and environmentally conscious citizens, the world remains imperfect. Nature herself fails to comply with many of the stringent standards that are judged by man to be desirable to protect the environment. Oil continues to naturally seep into the oceans in many localities throughout the world. Despite great debate over the disposition of dredged materials originaing from man's activities, the rivers of the world continue to build deltas into the ocean areas comprised of sediments derived from vast areas of the continents. Fresh water runoff during flood stages of rivers dilutes the salinity of sensitive bay and estuarine areas, thus devastating larval stages of shrimp and fish and causing temporary reductions in catches of these species. Trees emit photochemical oxidants into the atmosphere, causing heavy smog in certain areas of the continents.

"Regulatory action should take cognizance of the situation that exists in the real world surrounding us, and utilize a pragmatic approach toward regulating man's activities. For example, consider the issue relative to long-term effects of crude oil in the marine environment. Coal Oil Point is a natural oil seep occurring approximately five miles offshore from Santa Barbara, California. This seep is known to have existed throughout the earliest recorded history of California, and has likely existed for eons. Extensive studies have shown the ecosystem surrounding this natural seep to be normal and healthy. Similar studies around other natural seeps in the world have been made with similar results. These facts should be pragmatic evidence that there are no significant detrimental long-term effects from oil in the marine environment. This view is strengthened by the fact that all of the oceans of the world contain bacteria capable of assimilating and breaking down crude oil. The existence of these bacteria throughout the natural environment and their dependency upon oil as a nutrient suggest that oil is a natural part of the marine environment. Acceptance of this fact leads one to conclude that nature has accepted crude oil as a natural part of the marine ecosystem. Thus, searching for long-term effects would appear fruitless.

"Another topic under intensive debate concerning OCS operations is the discharge of drilling mud and cuttings into the marine environment. Despite intensive laboratory research and field monitoring tests which show the effects of drilling mud and cuttings to be relatively benign, controversy remains as to the toxic effects of chemicals contained in the mud, and the effect of sediment derived from mud and cuttings which are to be deposited upon the ocean floor. In the real world the Mississippi River alone deposits millions of pounds of sediments in the Gulf of Mexico each day, which is equivalent to the amount of sediments generated by thousands of wells. Mississippi River effluent contains chemicals in runoff from agricultural lands approximating two-thirds of the continent, in addition to those derived from manufacturing plants located adjacent to the river. By contrast chemicals normally found in drilling are of relatively low toxicity and monitoring tests have shown all effects from drilling mud are reduced by dispersion to an undetectable level within 100 meters of the point of discharge.

"Other interesting comparisions to the effects of drilling muds are those that may be derived from the fishing industry. It is interesting to note that one clam dredge on the Georges Bank disturbs and redistributes sediments in one day equivalent to drilling thirty-two 10,000-foot wells. Similar sediment disruption and redistribution is caused by all trawling operations conducted around the periphery of the United States on the OCS with total sediment disruption and redistribution each many thousands of times greater than that caused by the offshore oil industry. If the societal concern is the disruption of benthic communities on the ocean floor, one must ask why the total devastation of these communities by the heavy trawl boards and clamming operations of the fishing industry is a societally acceptable activity, when the discharge of a tiny fraction of sediments from OCS drilling activities is unacceptable.

"One final comment on pragmatism. There are those in our society who stand on the premise that no activity should be allowed until all facts are known. This premise has been used repeatedly regarding operations in the Outer Continental Shelf. When the preponderance of evidence indicates that no harm can be readily detected from an activity, critics back off to a position that further studies are needed to determine long-term impacts that may not be readily discernible from early research work. The truth is that mankind will never know all there is to know about the oceans and the marine environment. If we are to survive, and make reasonable use of the resources available to us, we must make practical decisions based on information reasonably obtainable. Activities causing demonstratable detrimental effects on the environment must be modified to overcome those impacts. Activities causing no significant detrimental effect on the environment must be allowed to proceed, leaving future generations with greater knowledge and better tools, to correct the errors made by this society." [Shirley]

INDUSTRY CAPABILITY AND PREPAREDNESS

C. L. Blackburn, Executive Vice President,
Shell Oil Company

"Now, I would like to review the impressive advances the industry has made during the '70s, and look at some of the offshore technology which will enable us to find and develop the needed resources during the 1980s.

"In the past decade, geophysical prospecting in the offshore has seen many advancements. New techniques enable our explorationists not only to accurately define the geometry of the subsurface rock layers, but also to predict the types of rocks and the presence of gas or oil in certain types of reserviors.

"Advances have been made in seismic data acquisition, processing and interpretation. Some of the latest developments in acquisition equipment are on display at this conference. New energy sources or arrays of more conventional sources provide a controlled signal several times stronger than the non-dynamite sources of ten years ago. The environmentally safe signal sent out from the survey vessel is continually measured and in most operations it is extremely consistent from shot to shot.

"Most of the processing advances in the early seventies were related to bright spot techniques which allow us to predict oil or gas from the seismic data. The list of new developments is a long one.

"Significant advances were made in:

- the control of the signal amplitudes to accurately quantify bright spots;
- the removal of multiples and noise;
- migration of dipping data into its proper two or three dimensional location; processing each trace into an approximation of a sonic log; and
- modeling to aid in the interpretation process.

"All of these and many other processes have become feasible because of the rapid advance in high speed, large capacity computers.

"Through advanced offshore seismic exploration, computer-controlled acquisition is monitored and pre-processed by technicians on board the vessel, speeding the data to large computer centers where processing and analysis is controlled through remote terminals. Teams of geophysicists, geologist, petrophysicists, and engineers interpret the data for subsurface structure. In addition, the interpretations today often provide vital information regarding rock characteristics, including fluid content, depositional environment, and drilling prognosis.

"The momentum of these technological advances can be expected to continue into the eighties, providing further improvement in industry's capability to locate reserves.

"Turning now to drilling, during the decade of the 1970s there was a dramatic increase in the industry's capability. In the early 1970s, offshore drilling activity in coastal waters of the U.S. averaged less than 100 rigs. . . .

"It was not until 1975 that rig activity climbed above that level as a result of greater recognition of the country's dependence on foreign oil and the strong control on that market by OPEC. Since that time, the growth in active offshore rigs has averaged 23 percent per year.

"An improved business environment will be essential to sustain this growth in the 1980s. There is no doubt that the offshore environment is getting more complex. Not only are the wells being drilled to greater depths, into higher pressure, and more corrosive environments, but such wells will also require operations in greater and greater water depths.

"Such operations are within the technological capability of the industry, but possible constraints on both technical and operations manpower, rig equipment, materials and capital funds are likely to moderate the growth rate compared with that of the late 1970s. Nevertheless, I expect the rig count to grow from the current level of 250 to about 400 active rigs in U.S. waters by 1990.

"Ten years ago, the industry was convinced that technology was available to drill and complete wells in water depths greater than 1500 feet and that 'economics not technology' would limit the progress into greater water depths. Those views proved prophetic in that the current water depth record of 4876 feet was achieved off Newfoundland in 1979.

"I am convinced that technology is currently available to permit the development of a system to drill and complete wells in 10,000 feet of water by the end of this decade. Since the offshore represents one of the most promising frontiers for primary oil and gas reserves, this province demands continued investment of both capital and manpower resources to develop this important U.S. asset.

"Technology advances in dynamic positioning of drilling vessels and improvements in riser designs and well control capability during the 1970s provide an excellent basis for future progress. Extension of this technology should be possible as the industry has the opportunity to gain more experience in deeper operations. Measurement-while-drilling instrumentation and telemetry should be developed during the 1980s to further improve the safety of well operations and the performance of rigs.

"I would now like to review offshore production systems. Nearly all offshore fields have been developed with fixed-

leg platforms. The number of drilling and production platforms in the Gulf of Mexico increased from 600 to 1132
during the 1970s. . . . There are now more than 1250 major
platforms in the Gulf, plus more than 2300 smaller well-
protector structure's. In addition, there are 22 platforms
and 7 islands installed offshore California, and 14 platforms in Cook Inlet, Alaska.

"Let's now talk about our ability to install platforms in
deep water. During the 1970s platforms were installed in
record water depths increasing from 373 feet in the Gulf of
Mexico, to 850 feet at the Hondo structure offshore California, to the current record of 1025 feet at Cognac in the
Gulf of Mexico. In the same period, large concrete platform
technology evolved in the North Sea with a record water
depth of 476 feet achieved.

"We believe that, technically, fixed-leg platforms can be
built for at least 1500 foot water-depth locations in the
Gulf of Mexico. However, due to the very large amount of
steel required and limitations of fabrication and installation methods, it is expected that the fixed-leg platform
will reach its economic limit at some shallower water depth.

"Two of the more promising concepts for extending the
water-depth capability are the guyed tower and tension-leg
platform. These are known as compliant structures which are
designed to move with the forces of wind, wave, and current
rather than ridgedly resist them.

"The guyed tower is a tall, slender structure that requires less steel than a fixed-leg platform. Guy lines are
used to hold the tower in its vertical position. Exxon has
successfully tested a one-fifth scale model in 300 feet of
water. They have announced plans to use a guyed tower to
develop a lease in 1000 feet of water in the Gulf of Mexico.

"Another platform type that holds considerable promise
for future use in deep water is the tension-leg platform. A
big advantage of the TLP is that steel requirements, and
thus cost, increase at a slower rate with water depth than
with either a fixed-leg platform or guyed tower.

"The other major type of deep-water production technology
involves subsea systems where wells are drilled from a
floating rig and completed on the sea floor. Production is
routed to surface facilities on a fixed or floating platform. About 200 wells have been completed on the sea floor
worldwide since 1960. There have been no significant pollution or accidents even though some subsea wells have produced for more than 16 years.

"Technical problem areas to be resolved by further development work are primarily related to hardware design that

will provide ease of installation, highly reliable operation, and efficient maintenance and repair. It is believed that subsea well and manifold equipment can be installed and operated in water depths to 10,000 feet, comparable to floating drilling depth capability.

"Offshore oil-storage and tanker-loading systems have not been used extensively in the United States. Only one system has been installed and that is Exxon's Hondo system offshore California. It is likely that some fields will be discovered, possibly off the Atlantic coast, where pipelines to shore may not be justified. Offshore tanker-loading systems will be required instead. Experience gained with such installations in other parts of the world indicates that technology is available to install and operate these systems in a safe and environmentally acceptable manner.

"This brings me to marine pipelines, where the industry made significant accomplishments in the decade of the 70s. A total of 4900 miles of pipelines were installed offshore the U.S., out of a total of 17,600 miles offshore worldwide.

"Development of deep-water pipelaying technology has consistently been ahead of field application. . . . The industry's dedication to pipeline research in the mid-'60s stimulated the development of mathematical models to analyze suspended pipe spans, plus such needed tools as the articulated stinger, buckle arrestors, and the semi-submersible-laybarge.

"Although lack of opportunity prevented early use of available pipeline capability on deepwater lines, the Brent gas line in the North Sea is a good example of the benefits of this technology. This line, 270 miles of 36-inch pipe in waters to about 550 feet, was laid trouble-free, ahead of schedule and under budget--quite an accomplishment for the North Sea.

"Following the success of the '70s, what is ahead for pipeline installations in the '80s? We envision no technical barriers to laying pipelines in water depths to 10,000 feet. Below about 3,000 feet, the installation method will probably change from today's horizontal stove pipe operation to a near-vertical initiation. The tendency will be to lay multiple smaller pipelines (16-20 inches) rather than large pipelines.

"In other underwater developments, the use of divers, remote vehicles, and mini-submarines for performing work underwater expanded significantly during the 1970s.

"Techniques for mixed-gas diving were perfected and have become commonplace. Saturation diving has been established as the accepted method for any extensive work deeper than about 300 feet.

"Working dives have been accomplished in water depths to 1200 feet, and simulated dives in controlled laboratory conditions have established that divers should be able to work with relative safety and efficiency at depths in excess of 2,000 feet.

"Remote-controlled vehicles have become an economical means for providing visual inspection of pipelines and platforms. In additon, some of the vehicles have manipulator capability for performing simple tasks.

"Mini-submarines have been used for pipeline and platform inspection work. The record water depth for oil industry use of mini-submarines is 4,876 feet in support of drilling operations. Tethered diving bells with thrusters and manipulators are also being used for drilling rig support work and have operated successfully in 2700 feet of water. These manned diving bells incorporate manipulator systems which have the sense of "feel," allowing the operator to perform very precise manipulative tasks.

"During the 1980s, all of these underwater work systems will be further developed to support offshore activities as they progress into deeper water.

"Finally, I want to spend a few minutes on the the technology we will be using increasingly in the 1980s and '90s in the frontier areas of Alaska.

"The Arctic offshore saw an increasing amount of activity in the 1970s. Alaskan activity, which had its start in Cook Inlet in the 1960s was primarily in the Beaufort Sea in the 1970s.

"The decade saw the development of temporary sand and gravel fill islands as safe, economical exploratory drilling platforms. Twenty-two gravel islands were built and used for drilling in the Canadian and U.S. Beaufort Sea during the years 1973 through 1980 in water depths ranging from four feet to more than sixty feet. Drilling operations have proceeded uninterrupted during the winter season on each of the islands. Wells have been also drilled from natural barrier islands and from a man-made ice island.

"In the Bering Sea, conventional jacket structures or ice-resistant concrete or steel towers may be used. In water depths of more than 600 feet, and possibly in some shallower locations, compliant structures and subsea completions such as those I described earlier may be used.

"Transportation will usually be by pipeline from the platform to shore, then to tankers or existing pipelines. Offshore storage and loading facilities may be used in deep-water and remote locations.

"Arctic offshore research and development has been underway for more than a decade. Ice research has concentrated on ice strength and stress-strain behavior, ice feature occurrence, ice movement, and ice structure interaction. Field programs to measure oceanographic parameters in the Arctic offshore have been underway for a number of years and will continue. Work will continue in evaluating platform concepts for the Arctic offshore.

"These technological developments are an important reason for the industry's excellent offshore safety record. Through the years, we have demonstrated that offshore wells can be drilled and produced in a safe, environmentally sound manner." [Blackburn, 1981]

PROBLEMS AND SOLUTIONS

B. John Mackin, Chairman, Zapata Corporation

"The offshore petroleum industry is no stranger to uncertainty. The political, strategic, and economic importance of oil as a crucial commodity has not made it easy for multinational companies in the private sector to make textbook business planning and investment decisions.

"Changes in government policy relating to various countries' positions as net importers of oil, or the size of the slice of the petroleum pie going to the home country, occur and we adapt to the new conditions, provided we are still left with an adequate return on our investment. Risk and unpredictability have come to be accepted as part of the game in the international offshore petroleum arena.

"What is not acceptable is risk and unpredictability in the domestic offshore petroleum arena caused by excessive government regulations, turf battles between the U.S. federal government and certain of the coastal states, and vexatious litigation brought by these same states and self-proclaimed environmental public interest groups. These forms of risk and unpredictability have nothing to do with the economic and political forces at work that make it challenging for petroleum companies to do business in the world today. On the contrary, such forces mandate that in our own best national interest we eliminate this conflict and unpredictability, get our domestic house in order, and get on with the business of developing our offshore oil and gas resources expeditiously.

"In 1980, the U.S. produced 760,000 barrels of oil per day--or just under 9 percent of domestic production--from the Outer Continental Shelf (OCS). This was down from 12 percent in 1971. In 1980, we also produced 12.5 billion cubic feet of natural gas per day from the OCS--or 23 percent of domestic production.

"In the first seven months of 1981 the OCS program generated $6.7 billion in revenue for the federal government, $5 billion in bonuses and $1.7 royalties. It is likely that revenues generated by the program for all of 1981 will top $10 billion.

"By most estimates, federal lands contain between one-third and one-half of our commercially recoverable reserves of oil and natural gas. The U.S. Department of the Interior estimates that 80 percent of the oil yet to be discovered on public lands in America lies beneath our OCS. Thus, as our nation strives for energy independence, it is imperative we concentrate more of our attention on the OCS and exploit these reserves.

"During the past 28 years, the federal government has leased some 22 million acres of offshore lands for oil and gas exploration after offering 40 million acres. Acreage offered for lease represents less than 3 percent of the total OCS acreage available for lease by the federal government. From this 3 percent oil companies have produced over nine billion barrels of oil and about 58 trillion cubic feet of natural gas. The federal government's cumulative revenue has been over $45 billion.

"Opponents of the program often overlook the fact that from the standpoint of benefits realized by the American public, this is one of the most successful programs in our history.

"During most of the 1970s the Interior Department leased an average of only slightly more than one million acres per year, while it was generally agreed that five to six million acres, onshore and offshore, had to be leased annually just to keep oil and gas production level.

"President Nixon gave greater recognition to the future potential of our OCS petroleum resources when he called for federal leasing of 10 million acres a year. Even with its protectionist disposition the Carter Administration called for leasing six to seven million acres a year.

"Given this background, it is easy to understand why the Reagan administration's proposed five-year offshore leasing program may, over the long term, turn out to be one of its most dramatic achievements. For the first time, the energy potential of America's offshore lands has been given Presidential recognition in the magnitude it deserves. During the five years from 1982 through 1986, the administration could offer for lease as much as 875 million acres located offshore Alaska, California and the Atlantic Coast, as well as in the Gulf of Mexico.

"This is a vitally important step forward for the United States. As reserves in the Gulf of Mexico and other known producing areas dwindle and become harder to find we must search the frontier areas included in the five-year plan for large reserves yet to be discovered. Like any responsible enterprise we must take inventory of all our potential resources so we can plan our future knowing all the relevant facts. Some of these acres will have deeper waters and more severe environmental conditions. However, industry has the willingness, technology, equipment, and manpower to do the job.

"Two factors remain, however, to obstruct the timely development of our OCS resources in the national interest. One is repeated vexatious litigation brought by certain of the coastal states and environmental special interest groups. The other is the maze of complex laws and regulations governing OCS leasing, exploration and production.

"The Santa Barbara oil spill in 1969 helped provide impetus to the environmental movement of the 1970s. Many federal laws were enacted during this decade which provided increased opportunities for coastal states to ensure the protection of their environmental interests. These included the National Environmental Policy Act of 1969, Marine Mammal Protection Act of 1972, Marine Protection Research and Sanctuaries Act of 1972, Coastal Zone Management Act of 1972, Endangered Species Act of 1973 and OCS Lands Act Amendments of 1978.

"During the same era, environmentally more restrictive amendments were also made to the Clean Air Act, the Fish and Wildlife Coordination Act, and the Federal Water Pollution Control Act among others.

"All of these laws have either directly granted coastal states dramatically increased participation in environemntal control over OCS oil and gas development, or have provided a legal basis for court challenges to federal management of the OCS initiated by states and so-called public interest groups. Indeed, as the Reagan administration took office, from the standpoint of both law and policy the pendulum had swung too far in favor of coastal state and protectionist interests.

"Coastal state governors are encouraged to submit comments relative to the five-year OCS leasing schedule required by the Act and to any proposed revision of it. The governors are entitled to a written response from the Interior

Secretary granting or denying any of their requests for modification of the five-year program and stating the reason for his decision.

"Likewise, regarding any governor's recommendation concerning the size, timing, or location of a proposed lease sale or a proposed development and production plan, the Interior Secretary shall accept such recommendations if he determines, after having provided the opportunity for consultation, that they provide for a reasonable balance between the national interest and the well-being of the citizens of the affected state. In order to strengthen the finality of such Secretarial decisions, the Congress provided that they shall not be subject to suit or judicial review unless found to be capricious.

"The OCS Lands Act Amendments of 1978 permitted and encouraged precedent-shattering state and local government participation in federal policy and planning decisions regarding OCS oil and gas activities by providing to coastal states extensive data and information; detailed advance notice of proposed leasing, exploration and development and production activities; and the procedural opportunity to comment and consult with the Secretary of the Interior on all such matters.

"As a general rule, states bordering the Gulf of Mexico, where offshore oil and gas development has been commonplace since the early 1950s, have indeed been supportive of the federal OCS program. Southern states bordering on the Atlantic Ocean have also lent support to the OCS program.

"On the other hand, California, Massachusetts, Maine, and Alaska in many instances have militantly resisted OCS oil and gas development. Each has argued for a number of different reasons that while offshore oil and gas resources should be developed, it would be preferable to develop such resources in OCS areas other than those located adjacent to their coasts.

"They have cited as the reason for their resistance to OCS leasing and related activities that the protection of other values should take precedence. These included protection of the marine environment--including fishery resources and marine mammals--tourism, and the establishment of marine sanctuaries, among others. Such arguments have questioned the validity of the federally endorsed multiple use policy which holds that the development of energy resources is quite compatible with the furtherance of other such values.

"The Coastal Zone Management Act provides that federal agencies conducting or supporting activities directly affecting the coastal zone shall conduct or support those activities in a manner which is, to the maximum extent practicable, consistent with approved state management programs.

"To suggest that a lease sale itself or any activity associated with it has a direct affect on the coastal zone is stretching logic and the plain meaning of the statute beyond the breaking point. Too many subsequent decisions involving discretion on the part of the leaseholder and federal and state agencies must be made before the coastal zone can be affected.

"The states have other opportunities to use their consistency club in virtually every phase of the OCS process. Under the 1978 OCS Lands Act Amendments, lessees and operators are required to submit an exploration plan prior to commencing exploratory activities under a lease, and the Department of the Interior must act on it within 30 days from submission. However, even though the plan may be approved within that time frame, no license or permit for any activity described in detail in such plan may be granted where an affected state has an approved coastal zone management plan unless the consistency requirements have been met.

"Upon discovery in a region other than the Gulf of Mexico, and prior to proceeding with development activity, a lessee or operator is required to file a development and production plan. If approval thereof does not require an environmental impact statement (EIS), any affected state (and local government) has 60 days to file an injunction.

"All through the 1970s coastal states, local governments and environmental groups teamed up to try to block OCS lease sales in the federal courts. In *Sierra Club* v. *Morton*, *County of Suffolk* v. *Andrus*, and *Alaska* v. *Andrus*, there was a repetitious pattern of unsuccessful plaintiff claims that pre-lease sale environmental impact statements were in violation of the National Environmental Policy Act of 1969 (NEPA). By the mid-1970s the federal agencies, as a result of having been subjected to intense NEPA litigation, had gained much experience in the preparation of environmental impact statements. Having suffered setbacks in their attempts to thwart OCS lease sales on NEPA grounds alone, coastal states joined with environmental plaintiffs to initiate further litigation. While continuing to rely on previously unsuccessful NEPA arguments, they suddenly began to turn to non-NEPA grounds upon which to attack OCS lease sales.

"The chronology of *Conservation Law Foundation* v. *Andrus* is long and drawn out. Georges Bank Sale 42˜ was originally scheduled for January 31, 1978. Less than two weeks before that date, the Commonwealth of Massachusetts and the Conservation Law Foundation of New England sought, and on January 28 obtained, a preliminary injunction against the sale. The U.S. District Court held, in part, that the Interior Secretary should have awaited enactment of the OCS legislation prior to his holding the sale.

"On September 28, 1978, President Carter signed the OCS Lands Act Amendments into law. Thirteen months after issuance of the preliminary injunction, i.e., on February 20, 1979, the First Circuit Court of Appeals reversed the District Court, holding that the waiting for enactment issue was moot, and remanded the case.

"Sale 42 was rescheduled for November 6, 1979. Plaintiffs again sought a preliminary injunction against the rescheduled Georges Bank sale, which was denied by the District Court, and the plaintiffs appealed. The first Circuit in *Conservation Law Foundation* v. *Andrus* sustained the District Court in an opinion handed down on November 6, 1979, the date on which the sale had been rescheduled.

"Although the sale was finally held on December 8, 1979, the plaintiffs managed to delay it for nearly two years. Approval of the Georges Bank exploratory drilling permits was delayed until June 29, 1981.

"In other vexatious and expensive litigation, California and Alaska have teamed up to fight the federal government's five-year OCS lease sale schedule. Also, in a case decided by the U.S. Supreme Court on December 1, 1981, the California State Lands Commission, the City of Long Beach, California and Energy Action Educational Foundation were defeated in their efforts to enjoin further OCS lease sales until the regulations regarding OCS bidding systems are changed.

"Earlier in 1981, the State of California and a coalition of environmental plaintiffs sued in federal court to enjoin the sale of 34 lease tracts offered in OCS Sale 53. The plaintiffs raised many of the same issues litigated in the OCS lease sale suits. For the first time, however, the Coastal Zone Management Act's federal consistency provisions have been raised as an issue. That case is now before the Circuit Court of Appeals.

"In July 1981, California filed another suit in federal court alleging that air quality regulations adopted by the U.S. Geological Survey for OCS areas adjacent to California are inadequate. California is seeking a court order which would apply more stringent local air quality regulations to OCS operations.

"All of these cases serve to illustrate unending coastal state efforts to thwart OCS oil and gas development in favor of local interests. Certainly coastal states are at no loss in ensuring that the marine environment adjacent to their coasts is being adequately protected. The question is whether in the process of protecting the coastal environment the states have too many weapons at their disposal.

"The business certainty that the authors of the 1978 OCS Lands Act Amendments said would follow that legislation has certainly not come about. Militant coastal states and environmental groups have cost American consumers and taxpayers billions of dollars in delayed bonuses and royalties, interest expense incurred by companies and inflated exploration and development costs. Not only has it been a challenge for oil company operators to plan and budget for lease sales and exploration and development activities, but it has also been difficult for service companies to plan their manpower needs and equipment expenditures in this stop again-start again atmosphere.

"Fortunately, the American oil executive has always had more than an average amount of entrepreneurship. For this reason companies have been willing to risk going forward in the face of uncertainty to make the necessary investments to assure there will be adequate technology, equipment, and manpower to get the job done. In the 1970s this was often costly to the companies and their stockholders.

"Recently environmental groups have claimed that industry does not have the capital, manpower, and equipment necessary to handle the Reagan administration's accelerated five-year leasing program. These groups are the last ones that should make this criticism, for the delays they have caused, more than any other factor, make it difficult to plan these necessary ingredients which are key to the success of the plan. But we will be ready despite them.

"A warning must be sounded, however. As we push farther and farther into the more extreme environments of areas included under the five-year plan, the technology and equipment will be expensive. Without more certainty in the process the entrepreneurial willingness of boards of directors to make decisions to invest $100 to $200 million in an offshore drilling rig will be tested.

"It is indeed ironic that, while the state bureaucrats and paid spokesmen for environmental special interest groups clamor for the best available and safest technology on the OCS, they have so little understanding of investment economics that they fail to comprehend that their own actions and the uncertainty they create actually threaten to retard the development of that technology.

"What can be done to alleviate these conditions and help bring about the more rational development of our offshore petroleum resources?

"First, a more realistic attitude on the part of coastal states and environmentalists which gives more recognition to the national interest and the safety of the OCS program.

"Secondly, regulations governing the OCS program developed during the Carter administration need to be streamlined and completely overhauled.

"Thirdly, while the reform and streamlining of regulations is vitally important to the future of OCS exploration and development, legislative reform is also needed. A recent study by the General Accounting Office (GAO) found that court challenges based on NEPA, ESA, and other statutes represent the greatest obstacles to timely development on the OCS and noted that such challenges have delayed operations from three months to two years. Unless ways are found to minimize such challenges, the GAO concluded that companies will never be assured they may engage in recovery activity on purchased leases.

"Federal revenue sharing has been suggested as a means of diffusing the states versus federal government conflict on the OCS, particularly as the Reagan administration phases down federal funding of state coastal zone management programs. The theory is that if states like California and Alaska receive a share of the bonuses and royalties paid for federal OCS activities off their coastlines, they would have a stronger economic incentive to support the federal program.

"Supporters also argue that the coastal states are the ones that feel the onshore impacts of the OCS program and bear a great deal of the risk for which they should be justly compensated. At the same time they say that revenue sharing is necessary if coastal states are to gain equity with inland states which already receive a share of revenue from minerals extracted from onshore federal lands.

"Critics of revenue sharing, on the other hand, argue that the coastal states received all they are entitled to from the Submerged Lands Act of 1953 which granted them title to the lands immediately adjacent to their coastlines. State lease sales have been held and thousands of wells have been drilled on state offshore lands, and the people of those states have benefitted handsomely from taxes, bonuses and royalties, critics add. The lands beyond the state boundaries are the nation's lands and they belong to the people of Iowa, Indiana, and Kansas as much as the people of California, Louisiana and Texas, it is argued.

"Revenue sharing will not be an easy issue to resolve, especially in the context of the drive for a balanced federal budget. Nonetheless, the idea has merit and it should be studied and debated further in the halls of Congress.

"As is evident along the U.S. Gulf Coast, coastal communities are impacted by federal offshore development, and, in all fairness, industry would do well to support a revenue sharing program which is part of much broader legislation intended to reduce the aforementioned impediments to OCS exploration and development.

"I cannot conclude without pointing out the special role which we in industry have to play, caught up as we are in this struggle between federal and state interests. Our duty is to be good stewards of the public lands entrusted to us as lessees and contractors and to be careful in our operations and considerate of state and local interests. We also have a responsibility to educate the public about both the energy potential and the safety of the OCS program. If we can do this successfully, we ourselves can help the OCS program proceed more expeditiously and predictably." [Mackin, 1981]

C.L. Blackburn

"Of the many factors that affect the future of the OCS, none is more important than the government framework in which we must operate. We seem reasonably assured of the presence of a significant resource base in the OCS which will provide the industry with many opportunities for discoveries. The technology necessary to do the work in deep water and other hostile areas is at hand and lacks only the opportunity for testing on actual problems in these areas to be proved. I am confident that, with some fine tuning by working with such problems in these environments, we have the tools to do the job.

"But the fact that we have the prospects and tools is not enough. We also must have a framework of administrative policies, legislation and regulation that allow us to operate.

"This framework must provide three things:

- timely and predictable access to the attractive areas in the OCS,

- a fair return on our investment of manpower and capital, and
- the opportunity to conduct our operations without unreasonable governmental constraint.

"Most importantly, this favorable governmental framework needs to have long term predictability, since the time from lease sale to production in some offshore areas may span two-to-four administrations.

"The on-again-off-again-lease schedules and energy policies of the past decade have contributed to the current and potential shortages of trained people and equipment.

"Long-term predictability is essential if we are going to train the people, expand our equipment-producing capability, and make the significant capital commitments that will be required in the '80s.

"I am confident that with a stable and predictable government framework, industry can and will make the commitments to take advantage of all opportunities.

"If we are to enjoy a favorable government framework for future operations in the OCS, we must work with this and future administrations and Congresses to:

- maintain a predictable leasing schedule to provide timely and efficient development of resources.
- eliminate constraints on our operations that produce no benefits for the public.
- restore the balance between needed exploration and development of the OCS and legitimate environmental concerns and . . .
- obtain through bidding systems and other financial considerations an equitable sharing of OCS revenues that will provide sufficient rewards for industry to encourage development and yet protect the public interst.

"We also must conduct our business and operations responsibly, both individually and as an industry.

"We must strive to make our operations even safer by proper training of our people and by eliminating shortcuts that increase risk. We must support responsible regulation of our activities as a means of better assuring uniform application of safe operating practices.

"Finally, we must work individually and collectively to inform the American public about the nature of OCS operations and the public benefits to be obtained from

development of the resources underlying the OCS. This will
be a most difficult task, but strong public support is our
best assurance of maintaining a governmental framework
favorable to the conduct of our future business.

"We have covered a lot of ground this afternoon, but I
think it provides the basis of the story we need to take to
the public.

- Adequate energy supplies are crucial to a healthy
  economy, and deepwater and arctic offshore areas
  offer the greatest potential for helping to meet
  those needs.
- Offshore technology and the people in this industry can find and produce the energy this country
  needs, safely and responsibly, if we are given the
  chance.
- Government policies and regulations can encourage
  offshore development, or government can prevent us
  from finding and developing the oil and gas this
  country needs.

"With increased opportunities to explore for resources
and an improved, cooperative, and consistent regulatory environment, I am confident this industry can fulfill the
promise that is being shown at this conference." [Blackburn,
1981]

REFERENCES

American Petroleum Institute. August 1981. "The Search for
    Offshore Oil and Gas--A National Imperative." Washing-
    ton, D.C.
Blackburn, C.L., "Offshore Oil and Gas Operations: Assess-
    ment of the '70s and Forecast for the '80s," Keynote Ad-
    dress, 1981 Offshore Technology Conference, General Ses-
    sion, Houston, Texas, May 4, 1981.
DiBona, Charles J., Remarks prepared for delivery before the
    Public Lands Conference, Denver, Colorado, November 19,
    1981.
Hay, Keith G., Conservation Director, American Petroleum In-
    stitute. Remarks at the 1979 Annual Meeting of the Amer-
    ican Association for the Advancement of Science, Houston,
    Texas, January 6, 1979.
Hester, Frank J., Private Consultant, and Evans, Robert, La
    Mer Bleu Productions, "Outer Continental Shelf Develop-
    ment in the Santa Barbara Channel: Lack of Detectable Im-

pact on Fishers." Offshore Technology Conference, OTC 2756, 9th Annual OTC in Houston, Texas, May 2-5, 1977.

Lassiter, Ronald C., "Georges Bank: Fish and Fuel," Proceedings from the Ninth Annual Sea Grant Lecture and Symposium, Kresge Auditorium, Massachusetts Institute of Technology, October 23, 1980.

Mackin, B., John, "Streamlining Offshore Development: A National Energy Priority," The International Essays for Business Decision Makers, vol. 6, The Center for International Business, Houston, Texas, December, 1981.

Marine Board, August, 1980. Committee on Assessment of Safety of OCS Activities, National Research Council, Assembly of Engineering.

McIntyre, A.D., Department of Agriculture and Fisheries for Scotland, Marine Laboratory, "Oil Pollution and Fisheries," Aberdeen.

Mertens, Edward W., Statement before the U.S. Department of Interior Bureau of Land Management Hearing on Proposed Oil and Gas Leasing on the South Atlantic Outer Continental Shelf (OCS Sale No. 43), Savannah, Georgia, and Charleston, South Carolina, March 28-April 1, 1977.

Shirley, O.J., "An Industry Perspective On Regulation Oil and Gas Operations Outer Continental Shelf," The Cost of Regulatory Compliance, Shell Oil Company, New Orleans, LA.

Watt, James G., Statement before the Subcommittee on Panama Canal/Outer Continental Shelf of the House Merchant Marine and Fisheries Committee, June 2, 1981a.

Watt, James G., "We Must Inventory Our Lands," Enterprise-- The Journal of Executive Action, July, 1981b: pp. 4-5.

Weiss, Dr. F.T., Statement of API before Committee on Commerce, Science and Transportation and Subcommittee on Energy Resources and Materials Production Committee on Energy and Natural Resources, United States Senate regarding Environmental Impact of Offshore Operations of the Petroleum Industry, Washington, D.C., December 5, 1979.

West, J. Robinson, Testimony before the Subcommittee on Panama Canal/Outer Continental Shelf, The House Committee on Merchant Marine & Fisheries, October 28, 1981.

8
Conclusion:
Landlords of the Sea

Joan Goldstein

> What we call land is an element of nature inextricably
> interwoven with man's institutions.
>
> Polanyi, 1944, 178

This book has been an in-depth study of one major U.S. energy policy centering on the search for oil and gas resources
beneath the sea. As we have seen, the controversies over
who actually owns the land beneath the sea has in the past
been and is currently the subject of intense concern for the
states of California, New England, the mid-Atlantic coastal
strip (as well as the South Atlantic), and Alaska.   In a
similar sense, on the international scale, the potentially
rich resources of offshore oil off the Falkland Islands'
coast is one underlying factor in the British-Argentine dispute.   The states, however, have resolved their disputes in
the court.   To understand these modern day domestic and international "sea wars," we must perceive them as land management and ownership issues, and, for the moment, overlook
the fact that this land happens to be located beneath the
sea.

The commodity value of (sea) land falls into three categories: one, the potential to yield resources; two, the
value of mineral rights; and, three, the potential to yield
economic activity.   For example, the state of New Jersey is
seeking out secondary industries, such as manufacturing contracts for pipeline materials and equipment, in anticipation
of the building of a sea-to-interior pipeline.   Almost simultaneously, across the Atlantic, the countries of England,
France, and Germany are vying for contracts to produce pipe

lines. The contractor? The Soviet Union. Here a discovery of considerable quantities of natural gas has regrouped old political alliances, bringing western and eastern Europe into new economic-political arrangements. But that is the subject of another book.

The U.S. "sea wars," which is the focus of this book, is a highly complex drama. We have presented all the critical points of view in the drama: and a drama it is when we consider that the cast includes 24 coastal states from as far north as Alaska in the West and as far south as Florida in the East; eight federal agencies and five divisions of the U.S. Department of the Interior; representatives of the fishing, oil, and gas industries; and environmental groups.

The sorting out of these complex organizations and their relationship to each other via policy committees, regional technical working groups, and scientific committees, would require nothing short of a libretto, one such as the kind that accompanies an opera in which there is a cast of thousands and where the audience does not speak the language of the text.

The offshore oil and gas policy program has acquired such magnitude because it is undoubtedly the most ambitious energy program of its kind from the point of view of government. It is ambitious in the sense that most or all of the potentially affected parties have gained access to the planning process. Yet, as this book points out, this access to the planning process is tempered by the question of who actually owns the land beneath the sea. As noted in an earlier book by this author (Goldstein, 1981), it is not the question of use versus nonuse that is at the base of natural-resource conflicts, but who manages the resource and therefore gains from that ownership.

In the case of the OCS (outer continental shelf) Lands Act of 1953, as amended in 1978, the major power struggle over the ownership of the land beneath the sea erupted between the coastal state governments, and the federal agencies. This happened because federal agencies act as power brokers for the government and the oil and gas industry in the sale of tracts of submerged land lying in the outer continental shelf. Under the OCS Lands Act, the Department of the Interior empowers their Bureau of Land Management to carry out leasing procedures, and collect the bonus money from the accepted high bids at lease sales. This bonus money then goes into the federal treasury. The other arm of the Department of the Interior, the U.S. Geological Survey, has the responsibility for monitoring drilling and production and for collecting royalties. Millions of dollars from offshore revenues are apportioned by Congress each year for distribution to states for parks and recreation projects and billions of dollars find their way into the federal treasury.

As we have read in this book, from the coastal state's perspective, there is a maldistribution of wealth and decison-making powers in the OCS policies. Colgan makes this point in the chapter on Georges Bank, and Kaplan in the chapter on California. Our authors' have pointed out that the critical turning point in that struggle between the federal and state governments came in the 1975 decision of the Supreme Court in the case of *United States* v. *Maine*. The hotly-disputed issue of who owns the outer continental shelf was resolved by the highest court in the land. The Supreme Court ruled that ownership of the continental shelf beyond three miles from shore was vested solely in the federal government.

Politically, this was a decision interpreted as a state's rights issue, and the states had lost. Economically, what this meant to the coastal states was that they would receive no gain whatsoever from the sale of land offshore beyond that three-mile limit and offshore drilling would certainly extend beyond that limit. Not only would the states fail to gain economically from the sale of the outer continental shelf to the oil and gas industry, but they could suffer the deleterious effects of oil spills and damage to the shoreline tourism industry.

From the federal and the industry perspective, on the other hand, the resolution of the six-year dispute between the federal government and the 13 Atlantic coastal states in the case of *United States* v. *Maine* had cleared a major legal hurdle for the stepped-up leasing process called for by the president. As noted in the chapters by Basille and the one by Shirley, the problems posed by states'-rights issues and environmental issues merely slowed up the process of a national energy recovery program.

The response of the states was to form their own organizations to maintain their preeminence in the decision-making process. Thus, as Wilson describes in his chapter, MAGCRC, the Middle Atlantic Governor's Coastal Resources Council was formed to offset the political control of the Department of the Interior. In California, Kaplan notes the formation by Governor Brown of his own state-level commission.

The policy issues that these MAGCRC, mid-Atlantic states addressed had to do with the question of authority and compensation. The council was to ask: Should the states have direct intervening authority in the OCS development decisions? The council also raised a policy issue that has yet to be resolved in 1982: Should the mid-Atlantic coastal states receive compensation for the costs due to OCS development and onshore impacts?

Meanwhile, the North-Atlantic states were beseiged by requests from the fishing industry to protect the industry and modify the Fishermen's Compensation Act. Compensation for economic and environmental damages to their shore-tourist industries and the fishing industry were prime issues for the Atlantic coastal states. The mid-Atlantic states and their northern New England neighbors, after all, were primarily older industrial regions with rapidly declining industrial systems. The escalating price of oil following the 1973-74 oil embargo had left both the mid- and North-Atlantic states badly scarred. They were industrial states in search of new industries and sources of revenue. The revenues from the lease sales would have nourished their economies. Moreover, their existing coastal industries were in jeopardy from oil spills, not only from the drilling, but potentially from the transportation of oil to refineries.

The response of the president to the state's pressure was to open up the planning process. In effect, Carter responded to one of the key policy issues raised by MAGCRC, that is, should the states have direct intervening authority. Thus, in 1979, the Intergovernmental Planning Program (IPP) was formed.

PARTICIPATION IN PLANNING POLICY

The Intergovernmental Planning Program for OCS Oil and Gas Leasing, Transportation, and Related Facilities, was established under the Bureau of Land Management. The six technical advisory committees covered Alaska, the Gulf of Mexico, the mid-Atlantic, the North Atlantic, the South Atlantic, and the Pacific regions. Moreover, the OCS Planning Program was to be overseen by a national policy committee, and augmented by a scientific committee.

The technical advisory committees were appointed by the secretary of the interior through the recommendations of the Bureau of Land Management and included representatives of each of the states, industry, related federal agencies, environmentalists, and on occasion, interested individuals. This author is an appointed member of this group as are most of the contributing authors of this book.

On paper, this appeared to be one of the most democratic, innovative, representative planning bodies of its kind.

Never before had an energy planning system engaged the participation of so many factions. Compared to the elite Nuclear Regulatory Commission, for example, the OCS program was a model of participatory democracy. In reality, however, the participation process for the states was restrictive. Despite the intricate and lengthy decision-making schema the states' participation was mainly to comment on the federal OCS process.

The one hook in the states' power of intervention was the consistency clause, an issue that is discussed at length in Kaplan's chapter on California and in Shirley's chapter. This issue will be elaborated on in the next section that deals with the changes generated by the Reagan administration.

CHANGES IN ADMINISTRATION: PRESIDENT REAGAN'S PLAN

As noted in this author's earlier book (Goldstein, 1981), nothing influences environmental decision making more than an election year and the subsequent change in administration. The election of a new president then, would initiate monumental policy shifts that the planning process and the planners must contend with.

The Reagan administration, with the appointment of James Watt as Secretary of the Interior, had formulated a new energy policy that included the acceleration of the rate and number of offshore drilling activities. To do this, from the state's perspective, Secretary Watt had created policies that were an attempt to divest the states of their role in decision making in offshore drilling.

One of these policies, as noted above, is discussed at length by Kaplan in her chapter on California. But briefly, Interior Secretary Watt moved to lease some 29 tracts in the California waters that had earlier been removed by former Interior Secretary Cecil Andrus after a vigorous battle with Governor Brown and several environmental groups. The Governor of California and his environmental constituents had sued Secretary Watt and won the case on the basis of a federal judgement. This judgement gave credence to the state's authority to prevent offshore drilling when it is found to be inconsistent with the state's federally-approved, coastal zone management program. This judgement, which is expected to be appealed, can only apply to those coastal states participating in the voluntary Coastal Zone Management Act. Not all the OCS affected states have chosen to participate in this program. California and New Jersey have done so; though New York and Virginia have not.

A second acceleration of the OCS process instigated by Secretary Watt involves the term "streamlining." This process is done by holding more lease sales in less time frames, thereby shortcutting the results of the Environmental Studies Program, a program that was noted as essential by Bert Brun in his chapter on environmental impacts.

Concern for the glossing over of environmental impact statements in advance of the drilling and exploration process has been expressed by various members of the Technical Advisory Committees. For example, at a recent meeting of one Atlantic committee, a state representative commented on the streamlining policy, "We have little enough input to re-sults--we have input, but not impact. We have had a change of philosophy in that procedure of including everything (in environmental impact statements)." Another representative, a geologist, was to comment, "Why is there no attention to geo hazards? Were they eliminated from consideration in the deletion options? It doesn't matter what you address--its what you intend to act upon. Get out all the energy you can, but in a safe and economic manner."

Finally, escalation of the conflict between the affected coastal states and the federal government will reach a peak during the summer of 1982. It is during that season that the Secretary of the Interior plans to offer for resale all of the tracts that had not been sold from 1980-82. For the mid-Atlantic alone, this could produce a lease sale of near-ly 300 tracts. Since these sales will reflect the stream-lining policy of Secretary Watt, the time relegated for en-vironmental impact statements and the thoroughness of that activity will be hastened considerably. The states, and particularly the Atlantic coastal states, will no doubt res-pond to this challenge to their decision-making powers.

The controversy over the new "streamlining" policies of Secretary Watt will be reflected nationwide. The acreage in each planning area included in Watt's new five year plan in-cludes: North Atlantic, 139 million (acres); the South At-lantic, 99 million; Western Gulf of Mexico, 35 million; Cen-tral Gulf of Mexico, 46 million; Eastern Gulf of Mexico, 58 million; Southern California, 22 million; Central and North-ern California, 37 million; North Aleutian Basin, 32 mil-lion; St. George Basin, 70 million; Navarin Basin, 37 mil-lion; Norton Basin, 25 million; Hope Basin, 12 million; Barrow Arch, 30 million; Diapir Field, 49 million; Gulf of Alaska, 133 million; Kodiak, 89 million; Cook Inlet,

8 million; and Shumagin, 84 million. The total for all planning areas is about one billion acres.

In commenting on his five-year plan for the outer continental shelf (OCS) oil and gas leasing program designed to promote offshore development, Secretary Watt made the following statement:

> In an effort to reduce America's dependence on foreign oil imports, the program will make over one billion acres of the OCS available for leasing within the next five years, give industry a broader choice in determining exploration strategies, and continue the excellent environmental record of the OCS program. [Watt: News Release, March 15, 1982]

New Reagan Policies: Parks and Oil

As we have stated before, the millions of dollars in revenues obtained from the governments lease sales of OCS lands to the oil and gas industries were entered into the U.S. Treasury. Their subsequent route to the Land and Water Conservation Fund was done for the purposes of acquiring new parkland and improving national and state parks in general. In February of 1982, the Reagan administration that had just asked for a cut of nearly $50 million in the budget for the National Parks Service, had also proposed to raise the fees that the government charges for use of the parks by tourists and recreationists. Thus, the park entrance fees and campground fees would be raised considerably, and so too would the cost of leisure for those citizens who utilize the parks.

In addition, the Land and Conservation Fund, a fund comprised of revenues raised from oil and gas royalties, has been the vehicle through which new park land was acquired. It was the proposal of Secretary of the Interior Watt to cease the use of these oil and gas revenues for the purchase of parklands, but to use some of the fund for park improvements, while putting the brakes on acquisition. Secretary Watt then took the energy and parks question a step further, and attempted to shift the policies in a manner that was viewed as alarming by the environmental community. Thus Watt proposed that public lands be drilled to develop new sources of oil, coal, and other energy resources. A discussion of this is noted in the Kaplan chapter.

Here we can observe that the revenues attained from the

federal lease sales to the oil and gas industries provide a substantial income for the federal government. The manner in which this income is distributed provides the federal administration with purchasing power to shift policies for example, on the acquisition of new national parks. Moreover, the federal agency could rescind the 80-year history of public lands as sacrosanct territory.

This example of policy formation serves to demonstrate the state governments' claim that the income from the OCS leasing process was a problem of maldistribution of wealth for those affected states. It also illustrates the intra-gency rivalries within the Department of the Interior. The Bureau of Land Management and the U.S. Geological Survey were engaged in revenue and land management; while the agencies of land regulation and protection, the U.S. Fish and Wildlife Service and the National Parks Service were cut back in budgets and decision-making roles. Not all the winners were on the federal level and all the losers on the state level. There was still one type of industrial gain for the OCS coastal states, if they could select that route, which many of them did not. That route, of course, is the development of onshore secondary industries from the activities of offshore drilling.

ONSHORE PIPELINES:
THE TRANSPORTATION OF OCS NATURAL GAS

The development of onshore activity connected with the transportation of anticipated discoveries of natural gas was the one economic incentive to the coastal states. Here, they anticipated they could create industries, expand the workforce, enlarge existing communities or develop entirely new ones.

The history of such onshore development met with mixed review. BLM manager, Esther Wunnicke, in her chapter on Alaska makes note of the care given by the companies to avoid creating the boom or bust syndrome so often documented in the old western mining communities. Wunnicke points out that the workers were lodged outside the community, particularly outside eskimo communities, and in that sense, little impact could be anticipated.

On the other hand, Mim Dixon's study of social impact assessment in Alaska, *What Happened to Fairbanks?* (1978) details the changes that can affect the stability of a community. Her conclusions were that planning and government assistance was necessary to offset the growth problems when a pipeline brings onshore development.

At the moment, New Jersey is considering the best routes to transport anticipated discoveries of natural gas from offshore New Jersey to existing onshore distribution facilities (New Jersey Department of Environmental Protection, 1980). Phase I of this study focuses on corridor identification. The focus of study is on potential damage to vegetation, terrestrial wildlife, and aquatic resources. The report looks at the impacts on communities only in passing suggesting that urban and suburban communities be ommitted as the sites onshore. But it is clear that the state government views the transportation plan as an economic boom. David Kinsey, director of the state Division of Coastal Resources has noted to this author that he sees no environmental problems with the proposed pipeline.

New Jersey is not Alaska, however. It is a highly urbanized region with unique rural areas, such as the Pine Barrens. It may be difficult to forecast the changes that such onshore development would entail. But pipeline development is at the moment the only means by which states attain economic benefits from offshore oil and gas drilling.

OFFSHORE PROBLEMS: SPILLS AND ACCIDENTS

The technology of deep-water drilling is new and relatively untried. There is widespread skepticism whether technology can be developed to drill adequately at 8,000 feet of very hostile water. Deep-water drilling is also very costly. It might well cost several hundred million dollars before anyone had the capability to bring oil out of those waters. The oil companies have commented that they would prefer to drill much closer to shore where the costs and risks are not so high.

The tragic accident on February 15 of the overturned oil rig in St. Johns, Newfoundland brought home the problems of drilling in hostile waters and the tragic loss of life.

The oil potential in the Hibernia field had been the economic hope of Newfoundland, and the means by which this poorest province in Canada could strengthen their economy. The price extracted for the new oil industry may have seemed too high for the Canadians who had lost family members on the rig, and who may face the problems of spills. It is not clear whether compensation is the proper redress for loss of life and land, but this too is a policy issue.

PROBLEMS AND POLICIES

We have addressed a number of social and economic problems that are associated with this relatively new form of energy recovery. We have noted that the planning process is broad based and comprehensive. Nevertheless, there are certain inequities that require the attention of policy makers. These problems range from the question of revenue sharing between the economic gains of the federal treasury and the affected coastal states to the impact of onshore pipeline development and to the environmental and economic costs of spills from tankers and accidents with rigs. In each case, there are economic and environmental considerations.

Theoretical Perspective

The great sociologist, Max Weber, has pointed out that "in the way in which the disposition over material property is distributed among a plurality of people, meeting competitively in the market for the purpose of exchange, in itself creates specific life changes" (Gerth and Mills, 1946, p. 181).

The "life chances" Weber refers to are opportunities afforded the owners as opposed to the nonowners in the competition for valued goods. Thus he points out that, "This mode of distribution gives to the propertied a monopoly on the possibility of transferring property from the sphere of use as a 'fortune,' to the sphere of capital goods. . . . Property and the 'lack of property' are, therefore, the basic categories of all class situations" (Gerth and Mills, 1946, p. 182). Weber defined differences between the "have" and the "have nots" as essentially a class struggle between social and economic opportunity structures, that application of theory was an industrial society in the early twentieth century. This book however, is examining the tensions between complex organizations in the late twentieth century. We are analyzing the relationship of multinational corporations to federal and state bureaucracies, but the theory is applicable, nevertheless. In this study the class system is a division between the benefits and gains of the federal government and the benefits and costs to the state governments. As such, the states have fewer life chances. Weber would concur. He notes that the determination of the class situation is by the market situation. "Classes are not communities," he states, "they merely represent possible and frequent bases for communal action" (Gerth and Mills, 1946, p. 181). It is property and the lack of property that are therefore the basic categories of all class situations.

The class situation as delineated by Weber is most clearly reflected in the relationship between the federal and state structures. The federal structure receives property and capital gain from the leasing of land beneath the sea, since this is property that is resold to the industry for oil and gas exploration. The states, on the other hand, are nonpropertied by virtue of the legitimate decision of the Supreme Court in the case of *United States* v. *Maine*. The states are, then, nonpropertied and have fewer life chances, that is, fewer opportunities in the economic sphere. The struggle between the coastal states and the federal system that is so thoroughly depicted in this book, can be defined as a Weberian "class situation." The struggle is between the propertied and the nonpropertied. Property in this case is the land beneath the sea, the outer continental shelf. How the nonpropertied attempt to rebalance the scales is the history of the OCS program. As such, there are a number of policy issues that need addressing.

POLICY ISSUES

Revenue sharing is one policy resolution of the states-federal conflict on OCS. This is a system by which states can receive economic gains from the recovery of oil and gas. It is understandable that such sharing of property gains would be a states' solution, not a federal one. Nevertheless, 1982 will undoubtedly be the year that revenue sharing will be discussed, if not resolved.

A second policy issue is the question of compensation for accidents. Some state's have drafted agreements with the federal government on damages to the tourist industry and to the environment. The question here is twofold. First, is the compensation offered sufficient to balance the decline of an older form of industry, such as tourism and fishing? The second part of that question deals with the compensating costs of cleanups when spills or other forms of environmental damages occur. In other words, who pays for the problems?

The third issue deals with decision-making power. We may ask, under the current administration, will the states' have even less power to comment on federal decisions?

The longer that these issues remain unresolved, the greater the intensity of power struggles between the two levels of government. At this point in time, energy policy formation with respect to offshore oil and natural gas centers on power struggles between levels of government and their mode of resolution.

SHIFTS IN INTERNATIONAL POLICIES

The international balance of power is shifting rapidly and may well affect the current U.S. policies on offshore oil leasing. The Middle East has lost their preeminance in the field of crude-oil production. The internal OPEC wars and the new development of offshore drilling in the China Sea, plus the greater effort toward consumer conservation have led to an oil "glut" in 1982 and the subsequent lowering of prices at the pump. These shifts in the economic and political balance of power must have their impact on the formulation of U.S. offshore energy policies. The less the marketplace enforces an energy crisis, the less the society is willing to accept the costs of offshore exploration and drilling. The benefits can only be valued when the crisis situation creates the pressure to accelerate the OCS process. Since there appears to be no immediate crisis of resource shortages, the "streamlining" policies of the current administration may well be under scrutiny by the coastal states and the country at large.

REFERENCES

Books:

Dixon, Mim. *What Happened to Fairbanks?* Boulder, Colo.: Westview Press, 1978.
Gerth and Mills. *From Max Weber.* New York: Oxford University Press, 1946.
Polanyi, Karl. *The Great Transformation.* Boston: Beacon Press, 1944.

Documents:

Testimony of Mayor John A. Markey, New Bedford, Mass., before the Ad Hoc Select House OCS Committee Oversight Hearing, March 1979.
Middle Atlantic Governors Coastal Resources Council. Identification & Analysis of Mid-Atlantic Onshore OCS Impacts. Delaware State Planning Office. Dover, Md.

U.S. Department of Interior, B.L.M. Draft Supplement to the
    Final Environmental Statement, Proposed Five Year OCS
    Oil and Gas Lease Sale Schedule.   January 1982-December
    1986. B.L.M.
U.S. Department of the Interior, Geological Survey and
    B.L.M.   Outer Continental Shelf Oil and Gas Information
    Program.   Atlantic Index (January 1975-November 1980)
    Rogers, Golden and Halpern.   Report 80-1202.

Newspapers:

New York *Times:* 1981.   November 11, 13, 14.
New York *Times:* 1982.   January 10, 19, 20.
New York *Times:* 1982.   February 6, 16, 18.
Home News (New Brunswick, N.J.): 1979, July 30.
Home News (New Brunswick, N.J.): 1980, March 6, 7, 12, 21.

About the Editor and Contributors

JOAN GOLDSTEIN is a sociologist with the Center for Human Environments, The Graduate Center, CUNY, New York. She is an appointed member of the Intergovernmental Planning Technical Working Committee on Offshore Oil Leasing, Outer Continental Shelf, Mid-Atlantic Region; and formerly was appointed to the Governor's Pinelands Review Committee. She is now a governor's appointed member of the New Jersey Public Health Council.

Dr. Goldstein is author of *Environmental Decision Making in Rural Locales: The Pine Barrens*, Praeger, 1981. She has delivered papers internationally at conferences in Europe and Canada.

Dr. Goldstein holds a B.A. from the University of Iowa, an M.S. from the Bank Street College, and a Ph.D. from the Graduate Center, City University of New York.

FRANK BASILE is manager of the Atlantic Coastal Region for the OCS, Bureau of Land Management, Department of the Interior. Mr. Basile has a B.A. from City College of New York and attended graduate programs at the University of Delaware.

BERT BRUN was formerly team leader with the U.S. Fish and Wildlife Services on OCS activities in the mid-Atlantic region. He is now with the Ministry of Agriculture and Fisheries in New Zealand in the Fisheries Management Division.

CHARLES S. COLGAN is senior economist and OCS program director for the State of Maine. Mr. Colgan has a B.A. from Colby College, and is a Ph.D. candidate from the University of Pennsylvania.

ELIZABETH R. KAPLAN is the legislative director of Friends of the Earth, Washington, D.C. She has a B.A. from Smith, and an M.A. from Middlebury College.

BARBARA KARLIN attended Vassar and received a B.A. from the University of North Carolina. She is public affairs officer for the New York OCS office of the Bureau of Land Management.

O.J. SHIRLEY is manager of Government and Industry affairs for the Shell Oil Company, Eastern Division in New Orleans. Mr. Shirley has a B.S. from the University of Oklahoma.

EDWARD WILSON is the OCS coordinator for the State of Virginia, chairperson of the Mid-Atlantic Technical Working Group, and member of the National Policy Committee. Mr. Wilson has a B.S. from Virginia Tech, and an M.S. from the Graduate School of Public Health at the University of Pittsburgh. He is an adjunct professor at the Hanneman Medical School in Philadelphia.

ESTHER C. WUNNICKE is the manager of the OCS activities of the Bureau of Land Management, Department of the Interior, Alaska. Ms. Wunnicke has an A.B. and a J.D. from George Washington University, and an M.Ed. from Adams State College in Colorado.

LIBRARY OF DAVIDSON COLLEGE

Books on regular loan may be checked out for **two weeks.** Books must be presented at the Circulation Desk in order to be renewed.

A fine is charged after date due.

Special books are subject to special regulations at the discretion of the library staff.